The Forms of Meaning

Approaches to Applied Semiotics
1

Mouton de Gruyter
Berlin · New York

The Forms of Meaning

Modeling Systems Theory and Semiotic Analysis

by
Thomas A. Sebeok
Marcel Danesi

Mouton de Gruyter
Berlin · New York 2000

Mouton de Gruyter (formerly Mouton, The Hague)
is a Division of Walter de Gruyter GmbH & Co. KG, Berlin

♾ Printed on acid-free paper which falls within the guidelines
of the ANSI to ensure permanence and durability.

Library of Congress Cataloging-in-Publication-Data

Sebeok, Thomas Albert, 1920–
The forms of meaning : modeling systems theory and semiotic analysis / by Thomas A. Sebeok, Marcel Danesi.
p. cm. – (Approaches to applied semiotics)
Includes bibliographical references and index.
ISBN 3 11 016751 4 (cloth : alk. paper) –
ISBN 3 11 016752 2 (pbk.: alk. paper)
1. Semiotics. 2. Communication models. 3. System theory. I. Danesi, Marcel, 1946– II. Title. III. Series.
P99.4.M63 S43 2000
302.2–dc21
99-058969

Die Deutsche Bibliothek – CIP-Einheitsaufnahme

Sebeok, Thomas A.:
The forms of meaning : modeling systems theory and semiotic analysis / by Thomas A. Sebeok; Marcel Danesi. – Berlin ; New York : Mouton de Gruyter, 2000
(Approaches to applied semiotics ; 1)
ISBN 3-11-016751-4 geb.
ISBN 3-11-016752-2 brosch.

Printing: W. Hildebrand, Berlin.
Binding: Lüderitz & Bauer, Berlin.
Cover design: Christopher Schneider, Berlin.
Printed in Germany.

Preface

One of the traits that distinguishes human beings from other species is an instinctive ability to make sophisticated, ingenious, resourceful *models*. Model-making typifies all aspects of human intellectual and social life. Before building a house, a constructor will make a miniature model of it and/or sketch out its structural features with the technique of blueprinting. An explorer will draft a map of the terrain s/he anticipates traversing. A scientist will draw a diagram of atoms and subatomic particles in order to get a "mental look" at their physical behavior. Miniature models, blueprints, maps, diagrams, and the like are so common that one hardly ever takes notice of their importance to human life; and even more rarely does one ever consider their *raison d'être* in the human species. Model-making constitutes a truly astonishing evolutionary attainment, without which it would be virtually impossible for modern humans to carry out their daily life routines. All this suggests the presence of a *modeling instinct* that is to human mental and social life what the physical instincts are to human biological life. Now, what is even more remarkable is that *modeling instincts* are observable in other species, as the relevant literature in biology and ethology has amply documented. The intriguing question that such deliberations invariably raise is the following one: *What is the function of modeling in life forms?* This question begs, in turn, a whole series of related ones: *How is human modeling similar to, or different from, modeling systems in other species? What is the relation between modeling and knowing?*

The purpose of this book is to present and describe a methodological framework that can be used to seek answers to questions such as these—a framework developed on the basis of the work that has been conducted in the field of inquiry known as *biosemiotics*. This is a movement within *semiotics* aiming to study the manifestation of modeling behaviors in and across all life forms. The framework is called *modeling systems theory* (MST), developed by one of the authors of this book—Thomas A. Sebeok—over a lifetime of research on the interface between the biological and the semiotic sciences (see, for instance, Sebeok 1994). This book is intended to be both a synthetic overview of MST and a compendium of illustrations showing how it can inform and potentially expand the method of inquiry in both semiotics and biology. Our principal goal is to distill from MST the main implications that we perceive it has for investigative practices in those sciences. Thus, we have written this book in an accessible "textbook" style, so that the reader can get a nontechnical, yet comprehensive, look at what MST is essentially all about.

This book is the result of a collaborative effort on two counts. First, it is the product of the research and ideas of the primary author, Thomas A. Sebeok, and of the practical implications that these have had for the secondary author, Marcel Danesi, in the courses he has been teaching in semiotic theory at the University of Toronto. Second, its specific layout and its contents have been guided by the many suggestions and commentaries that both colleagues and students, at Indiana University and the University of Toronto, have passed on to each author over the years. We sincerely hope that this book will reflect what they have told us would be most useful to them. This book can be used as a reference manual by semioticians interested in MST and by students taking advanced courses in semiotics, communication theory, media studies, biology, linguistics, or culture studies. We have composed it so that a broad readership can appreciate the fascinating and vital work going on in this relatively unknown area of inquiry, most of which is often too technical for general consumption. Each chapter contains numerous practical exemplifications and insights into the potential applications of MST to the study of cross-species modeling. Nevertheless, the writing is not so diluted as to make it an overly simplified treatment. Some effort to understand the contents of each chapter on the part of the reader will be required. The more technical parts might entail several re-readings.

Since the focus of this book is practical, the critical apparatus of references to the technical literature is kept to a minimum. For the sake of comprehensiveness, we have appended an extensive bibliography of works upon which the MST framework has been built at the back. A convenient glossary of technical terms is also included.

Contents

Chapter I
Models and semiotic theory

> *I can't work without a model. I won't say I turn my back on nature ruthlessly in order to turn a study into a picture, arranging the colors, enlarging and simplifying; but in the matter of form I am too afraid of departing from the possible and the true.*
>
> Vincent Van Gogh (1853–1890)

1. Introductory remarks

A striking feature of human cognitive and social activities is the fact that they are mediated by the innumerable *forms of meaning* created and conveyed by the words, drawings, artifacts and other *models* of the world that people make and use routinely. The world of human beings is a de facto world of *meaning-bearing forms*. The systematic study of these forms comes under the rubric of *semiotics*, defined commonly as the "science of signs".

Modeling is the innate ability to produce *forms* to stand for objects, events, feelings, actions, situations, and ideas perceived to have some meaning, purpose, or useful function. The form may be imagined, in which case it is called a *mental image*, or it may be something externalized, in which case it is called a *representation*. Semiotic research has identified four basic types of forms: (1) *signs* (words, gestures, etc.); (2) *texts* (stories, theories, etc.); (3) *codes* (language, music, etc.); and (4) *figural assemblages* (metaphors, metonyms, etc.). In this opening chapter, we will describe and illustrate each of these types, recasting their traditional conceptions in terms of an approach to semiotics known as *modeling systems theory* (MST). These "recastings" constitute the basic elements of a methodological framework, called *Systems Analysis* (SA), that can be applied to the study of modeling phenomena across species. The goal of SA is, in fact, to make the systematic study of cross-species modeling a practicable goal.

1.1 Models

What is a *model*? Although a model is easily recognizable as such, it is something that virtually defies a formal definition. As the philosopher Max Black pointed out in his classic (1962) study of modeling in science, the term *model* has as many definitions as it has uses. For the present purposes, a *model* can be defined as a *form* that has been imagined or made externally (through some physical medium) to stand for an object, event, feeling, etc., known as a *referent*, or for a class of similar (or related) objects, events, feelings, etc., known as a *referential domain*. An imagined form can be called, simply, a *mental form*; a form made externally to stand for a referent can be called an *externalized form*.

A toy model of a house, made with a set of plastic building blocks, is a perfect example of what constitutes a model. Clearly, it is an *externalized form*, since it has been constructed to *represent* the physical *form* of a real-world house–i.e., "present it again" in terms of the blocks. The house is the *referent* of this model. Needless to say, the degree of structural fidelity between the toy model and the actual house it attempts to duplicate will vary according to the specific abilities of the model-maker, the number and kinds of blocks available, and the degree of reproducibility of the house–if the house to be modeled has many architectural details, for example, then it is much harder to reproduce its form faithfully in the toy model.

Models serve many functions in human life. They allow people to recognize patterns in things; they act as predictive guides or plans for taking actions; they serve as exemplars of specific kinds of phenomena; and the list could go on and on. As mentioned, the science which studies models and their functions is semiotics. For the sake of historical accuracy, it should be mentioned that semiotics was founded as a branch of medicine in the ancient world. In fact, in its oldest usage, the term *semeiotics* was coined by Hippocrates (460?-377? BC), the founder of Western medicine, to designate the study of particular types of *forms*–bodily symptoms. Symptoms are, in effect, forms produced by Nature's own modeling systems, designed to alert an organism to the presence of altered states in its body. The particular forms that symptoms assume in a specific species provide vital clues to the probable source and etiology of such states.

A symptom is an example of an *externalized natural form*, i.e., a form produced by Nature. Words and symbols, on the other hand, are *externalized artificial forms*, i.e., forms made intentionally by human beings to represent something. There are four general types of artificial forms that humans are capable of producing: *singularized, composite, cohesive*, and *connective*.

In traditional semiotic theory *singularized forms* are called *signs*. In an MST framework, a sign can be defined, more precisely, as a form that has been made specifically to represent a singular (unitary) referent or referential domain. Singularized forms can be verbal or nonverbal. The English word *cat*, or the Spanish word *gato*, for example, are verbal singularized forms standing for the referent [carnivorous mammal with a tail, whiskers, and retractile claws]; figure 1 (below) is their nonverbal (visual) equivalent. (In this book, square brackets are used to enclose forms, referents, and features of various kinds). Now, a description of the same referent as *a popular household pet that is useful for killing mice and rats* constitutes, clearly, a different kind of form. This is known traditionally as a descriptive *text*. In MST, a text can be defined, more specifically, as a *composite form*; i.e., as a form that has been made to represent various referents–[household pet], [mice], etc.–in a composite (combinatory) manner. Classifying a *cat* in the same category as a *tiger, lion, jaguar, leopard, cheetah,* etc. exemplifies another type of modeling strategy–namely, the tendency to *codify* types of forms in some *cohesive* fashion. In MST, a code can be defined as a system that allows for the representation of referents perceived to share common traits–e.g., [cat], [tiger], [lion], [jaguar], etc. (= the feline code). Codes consist of interacting elements, forming a *cohesive* whole, which can be deployed to represent types of phenomena in specific ways. Finally, the use of the word *cat* in an expression like "Alexander is a cool *cat*" is the result of a fourth type of modeling strategy, known traditionally as *metaphorical.* In this book, the term *connective form* is preferred instead, because a metaphor is a form which results, in effect, from the linkage of different types of referents (or referential domains): e.g., a human referent, [Alexander], with a feline referent, [cat].

Figure 1. Nonverbal (visual) form standing for [carnivorous mammal with a tail, whiskers, and retractile claws]

Consider the toy house analogy again. Modeling a [house] with building blocks produces, in effect, a singularized form. However, if the construction were made to include pieces representing a surrounding lawn, fence, and road, then the model would exhibit a composite form. Now, if the same set of blocks could be used not only for making a specific kind of house form, but other habitation forms as well (a hut, a cabin, etc.), then that set of blocks would constitute a cohesive system, since it would allow for the modeling of different types of abodes. Lastly, if the set of building blocks designed for making model houses were augmented by a set of different kinds of building blocks–say, blocks designed for making model vehicles–then various new models could be envisaged: e.g., a mobile home, a house trailer, etc. These are connective forms, resulting from the linkage of different kinds of building blocks.

These four types of modeling strategies are not mutually exclusive. Indeed, they are highly interdependent–signs go into the make-up of texts which, in turn, are dependent upon the elements that codes make available. As an analogy, consider another type of toy–the jigsaw puzzle. In this puzzle, the following parallels can be made:

- *singularized form* = single piece of the puzzle
- *composite form* = the picture that results when the pieces have been assembled in the required manner
- *cohesive form* = the jigsaw puzzle itself as different from, say, a chess game

- *connective form* = any linkage made between the pieces of the jigsaw puzzle and those of chess

The last analogue is purely illustrative. Unlike the new forms resulting from metaphorical connections, a jigsaw puzzle piece linked with a chess piece in some way does not generate a new puzzle form.

The different artificial forms that characterize human representation are shown in figure 2:

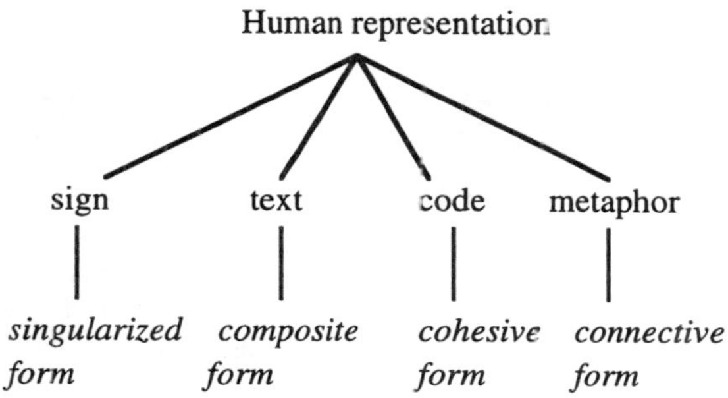

Figure 2. Types of forms characterizing human representation

1.1.1 Modeling, semiosis, and representation

The ability to make models is, actually, a derivative of *semiosis*, defined simply as the capacity of a species to produce and comprehend the specific types of models it requires for processing and codifying perceptual input in its own way. Semiosis is a capacity of all life forms; representation, on the other hand, is a unique capacity of the human species, which develops during the neonate and childhood periods. When an infant comes into contact with a new object, h/er instinctive reaction is to explore it with the *senses*, i.e., to handle it, taste it, smell it, listen to any sounds it makes, and visually observe its features. This exploratory phase of knowing the object constitutes a *sensory modeling* stage. The resulting internal model (mental image) allows the infant to *recognize* the same object subsequently without having, each time, to examine it over again "from scratch" with h/er sensory system (although the infant often will examine its physical qualities for various other reasons). Now, as the infant grows, s/he starts to engage more and more in semiosic behavior that replaces this sensory phase; i.e., s/he starts pointing to the object and/or imitating the sounds it makes, rather than just handling it, tasting it, etc. These imitations and indications are the child's first attempts at *representing* the world in purely human terms (Morris 1938, 1946). Thereafter, the child's repertoire of *representational* activities increases dramatically, as s/he learns more and more how to refer to the world through the singularized, composite, cohesive, and connective modeling resources to which s/he is exposed in cultural context.

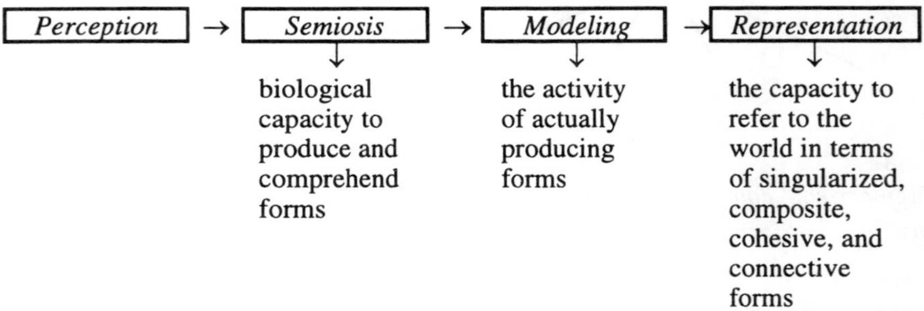

Figure 3. Relation among semiosis, modeling, and representation

1.1.2 Concepts

Representation reveals how the human brain carries out its work of transforming *sensory knowing* into *conceptual knowing*. Concepts are mental forms. There are two basic types of concepts–concrete and abstract. A *concrete concept* is a mental form whose external referent is demonstrable and observable in a direct way, whereas an *abstract concept* is a mental form whose external referent cannot be demonstrated or observed directly. So, for example, the word *car* stands for a concrete concept because its referent, [a self-propelled land vehicle, powered by an internal-combustion engine], can easily be demonstrated or observed in the physical world. The word *love*, on the other hand, represents an abstract concept because, although [love] exists as an emotional phenomenon, it cannot be demonstrated or observed directly, i.e., the emotion itself cannot be conceptualized apart from the behaviors, states of mind, etc. that it produces.

Concepts are formed in one of three general ways. The first is by *induction*. Induction is the process of deriving a concept from particular facts of instances. For example, a child who has not yet formed the concept of a [cat] might notice that certain types of animals s/he has encountered have whiskers. This feature would lead the child to induce that *any* such animal is imaginable (and thus representable) as a creature [with whiskers]. The second way in which humans form concepts is by *deduction*, the opposite of induction. For instance, a child who has formed the concept of [cat] would be able to deduce whether a specific mammal which s/he encounters for the first time is a [cat] or not by observing if it fits the general form of a [cat] in h/er mind. Finally, concepts are formed through *abduction*. For the present purposes, this can be defined simply as the process by which a new concept is derived on the basis of an existing concept which is perceived intuitively as having something in common with it. Abduction constitutes "best guess inferencing". A classic example of abductive reasoning is provided by the annals of science. The English physicist Ernest Rutherford (1871-1937) proposed a theory of atomic structure whereby he guessed that the inside of an atom had the structure of an infinitesimal solar system, with electrons behaving like little planets orbiting around an atomic nucleus. Rutherford's model of atomic structure was, in effect, an abduction as to what the inside of an atom looked like.

The distinction between concrete and abstract concept-formation is, needless to say, a convenient one. In actual fact, there are many degrees of concreteness and abstraction in conceptualization that are influenced by various kinds of psychological and social factors (Leech 1981: 9-23). Suffice it to say here that most of the raw, unorganized information that comes from seeing, hearing, and the other senses is organized into useful *concepts* by representational forms that have been arrived at through induction, deduction, or abduction. Moreover, it is now evident that the type of conceptualization process enlisted depends on the kind of form that the human mind seeks to extract from a specific situation. Often, all three processes–induction, deduction, abduction–are involved in a complementary fashion.

Since concepts are mental forms, it follows that the *form* that *knowledge* assumes depends on the type of *modeling* used. To see why this is necessarily so, consider the following anecdotal rendition of the notion of *indeterminacy* in physics formulated by Werner Heisenberg (1901-1976). Suppose that a scientist reared and trained in New York observes a physical event that s/he has never seen before. Curious about what it is, s/he takes out a notebook and writes down h/er observations in American English. At the instant that the American scientist observes the event, another scientist, reared and trained in the Philippines who speaks only the indigenous Tagalog language, also sees the same event. The Philippine scientist similarly takes out a notebook and writes down h/er own observations in Tagalog. Now, the question is: *To what extent will the contents of the observations, as written in the two notebooks, coincide?* The answer, of course, is that the two sets of observations will not coincide completely. The reason for this is not due, clearly, to the nature of the event, but rather to the fact that the observers were different people, and that the representational systems used (English vs. Tagalog) provided each scientist with different verbal forms for characterizing the event. So, as Heisenberg aptly suggested, the true nature of the event is indeterminable, although it can be investigated further, paradoxically, on the basis of the notes taken by the two scientists. Those notes are the de facto *models* that the scientists made of the event, both of which can be used to conceptualize the event, albeit *from different representational perspectives.*

1.1.3 Forms of meaning

The psychologist C. K. Ogden and the literary critic I. A. Richards argued in their classic 1923 work, titled appropriately *The Meaning of Meaning*, that it is impossible to define the notion of *meaning*. To the best of our knowledge, no significant progress has been made since in defining this term with any degree of accuracy. For the present purposes, it is sufficient to equate *meaning* with the particular *concept* elicited by a specific *representational form.* In traditional sign theory, the former is called the *signified*, and the latter, the *signifier.*

Human representation, as Ogden and Richards further observed, is a highly variable process. Like the indeterminacy involved in understanding natural phenomena, so too the exact nature of a *signified* is indeterminable in any objective sense, because its interpretation is shaped by situation, context, historical processes, and various other factors external to semiosis.

Semiotic theory has identified three main kinds of signifieds that representational forms encompass: *denotata, connotata*, and *annotata*. A *denotatum* is the initial referent (or referential domain) captured by a form. The process of representation in this case is called *denotation*. The denotatum of the word *house*, for example, elicits a singularized referent, namely [structure for human habitation]. Now, in human representational behavior, denotata can be extended freely to encompass other kinds of referents or referential domains, known as *connotata*, that appear to have something in common with them. This extensional process is known as *connotation*. For example, the meaning of *house* as a [structure for human habitation] can be extended to encompass connotata such as [audience], as in "The *house* roared with laughter"; and [legislative assembly], as in "The *house* is in session now". The salient characteristic of such connotata is that they extend the form of the initial referent–e.g., [structure for human habitation]–by implication: audiences and legislative assemblies do indeed imply [structures] of special kinds that [humans] can be said to [inhabit (occupy)] in some specific way. Texts and codes can likewise be extended freely to encompass an infinitude of connotata. Dress codes, for example, are regularly designed to evoke diverse social and/or group-specific connotata. Finally, the meaning of any form is influenced by subjective and/or group-specific interpretive *annotata*: e.g., the word *house* elicits subjective meanings that will vary according to an individual's or specific group's perception of [structures for human habitation]. An *annotatum* can be defined simply as the interpolation or assignment of a subjective and/or social meaning to a form (sign, text, etc.).

1.2 Modeling systems

The types of forms discussed above are the end-results of representational activities undergirded by three different, but interrelated, *modeling systems* present in the human brain, corresponding *grosso modo* to what Charles Peirce (1839-1914) called *firstness*, *secondness*, and *thirdness*. The child's earliest strategy for knowing an object with h/er senses is, in fact, a *firstness* strategy (see above §1.1.1). The modeling system that underlies firstness forms of representation is the *primary modeling system* (PMS). The PMS can be defined as the instinctive ability to model the *sensory* or *perceptual* properties of referents. The child's subsequent attempts to refer to the object through vocal imitation and/or manual indication constitute a *secondness* knowing strategy. The modeling system that guides these attempts is the *secondary modeling system* (SMS). The SMS can be defined as the capacity to refer to objects with *extended* primary forms and with indexical (indicational) forms. Finally, in learning to use a culture-specific name to refer to an object, the child is engaging in a *thirdness* form of knowing. H/er ability to do so is dependent upon the *tertiary modeling system* (TMS), which can be defined as the capacity to acquire and utilize the symbolic resources of culture-specific abstract systems of representation.

These three systems can be characterized succinctly in developmental terms as follows:

- *Primary Modeling System* (PMS) = the system that predisposes the human infant to engage in sense-based forms of modeling.
- *Secondary Modeling System* (SMS) = the system that subsequently impels the child to engage in extensional and indexical forms of modeling.
- *Tertiary Modeling System* (TMS) = the system that allows the maturing child to engage in highly abstract (symbol-based) forms of modeling.

1.2.1 Modeling systems theory

Although modeling systems theory (MST) has roots in the work of various twentieth-century structuralist semioticians, it has never really blossomed forth as a comprehensive theoretical and methodological framework for general use in theoretical semiotics (e.g., Sebeok 1994). The elemental axiom, around which we have fashioned our own framework for MST in this book, is the conception that all representational phenomena can be grouped into four broad types–singularized, composite, cohesive, connective. From this axiom six principles follow:

- Representation is the end-result of modeling (the *modeling principle*).
- Knowledge is indistinguishable from how it is represented (the *representational principle*).
- Modeling unfolds on three levels or dimensions, of which iconicity and indexicality (see below §1.3.2) are prior developmentally and cognitively to symbolicity (§1.3.2) (the *dimensionality principle*).
- Complex (abstract) models are derivatives of simpler (more concrete) ones (the *extensionality principle*).
- Models and their meanings are interconnected to each other (the *interconnectedness principle*).
- All models display the same pattern of structural properties (the *structuralist principle*).

Needless to say, we cannot go here into the many interesting philosophical problems related to what is knowledge. The *representational principle* implies simply that in order for something to be known and remembered, it must be assigned some representational form. The *modeling principle* asserts that modeling is the activity that underlies representation. The *dimensionality principle* maintains that there are three dimensions or systems involved in modeling–primary (iconicity), secondary (indexicality and extensionality), and tertiary (symbolicity). The *extensionality principle* posits that abstract forms are derivatives of more concrete, sense-based forms. The *interconnectedness principle* asserts that a specific form is interconnected to other forms (words to gestures, diagrams to metaphors, etc.). The *structuralist principle* claims that certain elemental structural properties characterize all modeling systems and forms. These are: *paradigmaticity, syntagmaticity, analogy, synchronicity, diachronicity,* and *signification.*

1.2.2 Structural properties

Paradigmaticity is a minimal differentiation property. To speakers of English, the two words *pin* and *bin* are kept distinct by a perceptible auditory difference in their initial sounds. This differentiation feature of sound systems is known in linguistics as *phonemic opposition*. Similarly, in classical Western music, a major chord is perceivable as distinct from a minor chord in the same key by virtue of a half tone difference in the middle tone of the chord. As such examples show, paradigmaticity is definable as the property of forms whereby some minimal feature is sufficient to keep them differentiated from all other forms of the same kind.

Syntagmaticity is a combinatory property. Forms such as *tpin, tpill, tpit,* and *tpeak*, for instance, would not be legitimate words in English because the initial sequence /tp/ + [vowel] is not characteristic of English word-formation, whereas words beginning with /sp/ + [vowel] would: *spin, spill, spit, speak*. This combinatory feature of words is called *syllable structure*. Similarly, a major chord is recognizable as such only if the three tones are combined in a specific way: [tonic] + [median] + [dominant]. Syntagmaticity is definable as the property whereby the components of a form are combinable in some specifiable way.

Analogy is an equivalence property, by which one type of form can be replaced by another that is perceived as being comparable to it. The English word *cat* is analogous to the Spanish word *gato*; European playing cards can replace American cards if an analogy is made between European and American suits; Roman numerals can replace Arabic numerals through simple conversion; and so on.

Synchronicity refers to the fact that forms are constructed at a given point in time for some particular purpose or function; and *diachronicity* to the fact that they are subject to change over time. The change that a form undergoes is not random, but rather, governed by both structural tendencies characterizing the code to which it belongs and external contextual (social, situational, etc.) influences. As an example, consider the word *occhio* 'eye' in Italian. The original form of this word was Latin *oculus*. Over time it became *oclu* (as various philological sources attest), and then *occhio*. These changes in physical form, however, did not come about haphazardly.

The elimination of the middle vowel of *oculus (oclus)* and the subsequent change of *cl* to *cchi* (phonetically [kky]) were structural tendencies within the late (Vulgar) Latin phonemic system.

Finally, *signification*, refers to the relation that is established between a form and its meaning. It is, more strictly, the relation that holds between the physical make-up of the form itself, the *signifier*, and the *referent* or *referential domain* to which it calls attention, namely the *signified*. As we saw above, there are three kinds of signification processes (§1.1.3)–denotation, connotation, and annotation.

The structural properties of forms are summarized below in table 1:

Table 1. Structural properties of forms

Property	Features/Functions/Manifestations
paradigmaticity	differentiation, recognizability
syntagmaticity	combination, arrangement
analogy	equivalence, replacement
synchronicity	structure and meaning of a form at a specific point in time
diachronicity	change in the structure and/or meaning of a form over time
signification	denotation, connotation, annotation

1.2.3 Biosemiotics

The modern-day practice of semiotics traces its origins to the writings of two scholars at the threshold of the twentieth century–the Swiss linguist Ferdinand de Saussure (1857-1913) and the American philosopher Charles S. Peirce (1839-1914). As an autonomous field of inquiry, it was expanded and developed throughout the twentieth century by Charles Morris, Roland Barthes, Louis Hjelmslev, Roman Jakobson, A. J. Greimas, Claude Lévi-Strauss, Juri Lotman, Thomas A. Sebeok, and Umberto Eco, to mention but a few.

As mentioned above (§1.1), in its oldest usage, the term *semeiotics* was coined by Hippocrates to alert medical practitioners to the value of knowing how to decipher bodily symptoms in order to carry out accurate diagnoses and formulate suitable prognoses of diseases. The study of *sema* 'signs' became the prerogative of philosophers around the time of Aristotle (384-322 BC) and the Stoic philosophers who were, in fact, among the first to take on the task of investigating word-signs in non-medical terms, characterizing them in terms of three dimensions: (1) the physical word itself (e.g., the sounds that make up the word *blue*); (2) the *referent* to which it calls attention (a certain category of color); (3) its evocation of a *meaning* (what the color entails psychologically and socially).

The next major step forward in the study of forms was the one taken by St. Augustine (354-430 AD). This philosopher and religious thinker was among the first to distinguish clearly between *natural* and *conventional* (artificial) forms, and to espouse the view that there was an inbuilt *interpretive* component to the whole process of representation–a view that was consistent with the hermeneutic tradition established by Clement of Alexandria (150?-215? AD), the Greek theologian and early Father of the Church.

John Locke (1632-1704), the English philosopher who set out the principles of empiricism, introduced the formal study of signs into philosophy in his *Essay Concerning Human Understanding* (1690), anticipating that it would allow philosophers to understand the intrinsic relation between representation and knowledge. But the task he laid out remained in virtual disregard until the writings of Saussure and Peirce. It is the work of the latter two which contains the foundational concepts for circumscribing an autonomous field of semiotic inquiry, aiming to study signs as elements related to each other systematically, rather than as isolated, material things in themselves. The key concept in semiotics is, in fact, that no single form can bear meaning unless it enters into systematic connections with other forms. These connections are traditionally considered to be binary in nature. Recall from above (§1.2.2) that the words *pin* and *bin* are kept distinct by a perceptible auditory difference in their initial sounds. This paradigmatic feature is the result, in effect, of a *binary opposition* between initial /p/ and /b/: the former is a voiceless consonant (produced without the vibration of the vocal cords); the latter a corresponding voiced consonant (produced with the vibration of the vocal cords). The physical feature [vibration of the cords], more commonly designated as [voice], is either present [+] or absent [-] in the constitution of a sound. Structurally, [±voice] is a binary phonetic feature that keeps various sounds distinct. The initial sounds in *pin* and in *bin* are two such sounds: /p/ is articulated as [-voice] and /b/ as [+voice].

MST is one of the fruits of an evolutionary branch of semiotics that has come to be called *biosemiotics* (e.g., Sebeok and Umiker-Sebeok 1992; Hoffmeyer 1996). The aim of biosemiotics is to extend the notions of general semiotics to encompass the study of semiosis and modeling in all species. The premise which guides biosemiotics is, in fact, that the forms produced by a specific species are constrained by the *modeling system(s)* which has evolved from its anatomical constitution. The aim of biosemiotics is to study not only the species belonging to one of the five kingdoms, Monera, Protoctista, Animalia, Plantae, and Fungi, but also to their hierarchically developed component parts, beginning with the cell, the minimal semiosic unit, estimated to consist of about fifty genes, or about one thousand billion (10^{12}) intricately organized atoms. Viruses are omitted from the biosemiotic purview because they are neither cells nor aggregations of cells.

Human bodies are assemblages of about one hundred thousand billion (10^{14}) cells, interconnected by an incessant flux of vital nerve signals. The origin of nucleated cells lies in a "semiosic collaboration" among single cells, which evolved less than one billion years after the formation of Earth. Simple cells likely fused at a certain point in time to form the complex assemblages of cells composing each living being. These assemblages constitute organs, which, in turn, constitute organisms, and which, in their turn, lead to the constitution of social systems (interacting organisms) of ever increasing complexity. The genetic code, of course, governs the exchange of signals on the cellular level; hormones and neurotransmitters mediate among organs and between one another (the immune defense system and the central nervous system are interconnected by a dense flow of two-way signal traffic); and a variety of signals conjoin organisms into a network of relations with each other as well as with the environment which sustains them.

In a phrase, the target of biosemiotics is the semiosic behavior of all living things. The main branches of biosemiotics are *phytosemiotics*, the study of semiosis in flora (Krampen 1981), *zoosemiotics*, the study of semiosis in fauna (e.g., Sebeok 1963, 1972a), and *anthroposemiotics*, the study of semiosis in humans:

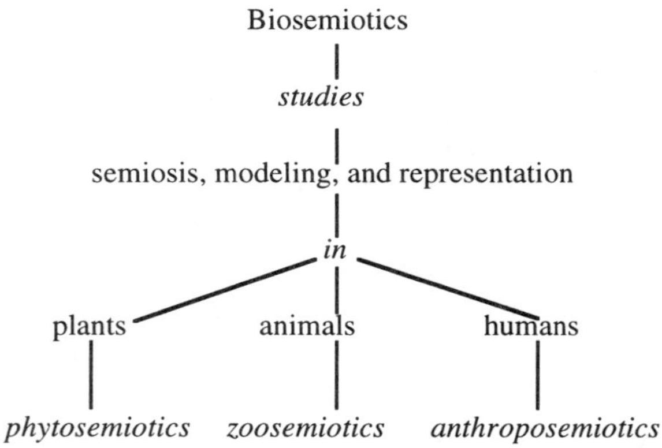

Figure 4. Branches of biosemiotics

In general, the method of inquiry in zoosemiotics is differentiated according to whether the animal is a herbivore or a predator, since the nutritional mode of the animal species shapes the features of its modeling system. The study of anthroposemiosis requires special treatment because the most distinctive trait of human semiosis is that it permits both nonverbal–demonstrably derived from its primate ancestry–and uniquely verbal modeling. The study of verbal modeling behavior constitutes the subject matter of the most advanced and highly formalized branch of semiotics, *general linguistics*.

Semiosis occurs at a molecular and chemical level first and is, thus, regulated by the genetic code, by humoral as well as cell-mediated immune reactions, and by the large number of peptides present in the central nervous system, functioning as neurotransmitters. The olfactory and gustatory senses are likewise "semiochemical". Even in vision, the impact of photons on the retina differentially affects the capacity of the pigment rhodopsin, which fills the ocular rods to absorb light of different wavelengths. Acoustic vibrations and tactile impulses delivered via the thermal senses are also transformed into electrochemical signals. Such signaling systems are routinely linked by several channels simultaneously or in parallel–a linkage that introduces a degree of redundancy, by virtue of which it becomes more likely that errors in reception will be minimized.

The founder of the biosemiotic movement was the Estonian-born, German biologist Jakob von Uexküll (1864-1944) who was among the first to document manifestations of different types of semiosic behaviors exhibited by different phyla. The crux of von Uexküll's approach is his contention that every organism has different inward and outward modeling strategies for monitoring information, which correspond to what we have called above imagistic and externalized modeling (§1.0). The key to understanding this duality is in the anatomical structure of the species itself. Animals with widely divergent anatomies do not undergo the same kinds of experiential processes, and thus possess vastly different modeling systems for monitoring such processes. Von Uexküll argued that an organism does not perceive an object in itself, but according to its own particular kind of modeling system(s), which allows it to interpret and model the world in its own peculiar biologically-programmed way. There exists, therefore, no analogous modeling system shared by humans and animals equally for encoding experiences in memory for future utilization.

The biosemiotic movement aims to investigate how all living things are endowed genetically with the capacity to use modeling for survival, and how anthroposemiosis is both similar to, and different from, phyto- and zoosemiosis, by establishing a taxonomy of notions, principles, and procedures for understanding the uniqueness of human modeling phenomena.

The study of zoosemiosis and animal communication actually traces its roots to Darwinian evolutionary biology (Darwin 1859), and especially to Darwin's (1872) claim that animal behavior constituted a viable analogue for human behavior. By the end of the nineteenth century, Darwinian-inspired work led to the establishment of comparative psychology. Some of the early animal experiments within this movement led to the theory of classical conditioning in humans. This was formulated originally by Ivan Pavlov (1902), following his well-known experiments with dogs, which showed that dogs could be conditioned to salivate at the sound of a bell. Initially, ringing a bell does not evoke a salivation response in a dog. So, Pavlov presented a meat stimulus to a dog while he rang a bell. After repeated bell ringings in tandem with the stimulus, Pavlov found that the bell alone would evoke salivation. The dog had been obviously conditioned (reprogrammed) to associate the sound of the bell to the presence of meat. Intrigued by such findings, the early psychologists devised ingenious experiments on animal behavior during the first quarter of the twentieth century. Robert Yerkes (1916), for instance, succeeded in showing that apes had the capacity to transfer their conditioned responses to novel learning tasks. And in 1925 Wolfgang Köhler found that simians could invent clever solutions to problems without previous training.

The goal of the early psychologists was not, however, to study animal behavior in itself, but to generalize the findings from the animal experiments to human behavior. The assumption was that the same "laws of behavior" applied across all species and, therefore, that universal principles of learning and problem-solving could be inferred from specific animal behaviors. By the mid part of the twentieth century, the use of animals as convenient substitutes for people in the laboratory came under attack and a new movement emerged, known as *ethology*, which stressed that animals and people lived in separate behavioral worlds, and that animals should be studied within their natural habitats, not in laboratories. The ethological movement came forward to establish the basic research techniques and categories for studying animal behavior in its own right.

In the 1950s and 1960s, linguists and semioticians came to regard the study of animal communication as particularly relevant to their own fields of inquiry. A slew of widely-popularized (some still ongoing) primate language projects were initiated, catching the attention of scientists and the general public alike. These were motivated initially by the proposition that interspecies communication was a realizable goal. However, although there have been reports of some sophisticated verbal activity and of some comprehension of humor, the ape experiments have not as yet established that primates (other than humans) have the capacity to full human language.

Since gorillas and chimpanzees are incapable of articulated speech, because they lack the requisite vocal organs, the first experimenters chose American Sign Language (ASL) as the teaching code. One of the first widely-known subjects was a female chimpanzee named Washoe whose linguistic training, by the Gardner husband and wife team (Gardner and Gardner 1969, 1975), began in 1966 when she was almost one year of age. Remarkably, Washoe learned to use 132 ASL signs in just over 4 years. What was even more striking was Washoe's acquired ability to put signs together to express a small set of composite messages.

Inspired by the results obtained by the Gardners, several others embarked upon an intensive research program aimed at expanding upon their teaching procedures. The Premacks (e.g., Premack and Premack 1983), for example, whose work actually began as far back as 1954, taught a chimpanzee named Sarah a form of written language, instructing her to arrange and respond to vertical sequences of plastic tokens on a magnetic board representing individual words: e.g., a small pink square = [banana]; a small blue triangle = [apple]; etc. Sarah apparently developed the ability to respond to combinations of such symbols, which included references to abstract notions. In the 1970s, Penny Patterson, while a Ph.D. candidate at Stanford University, taught a gorilla named Koko to use sign language. Amazingly, Koko eventually asked her trainer for a voice so that she could speak like her (Patterson 1978).

Although there continues to be enthusiasm over such results, with the media reporting on them on a regular basis, there really has emerged no solid evidence to suggest that chimpanzees and gorillas are capable of verbal behavior *in the same way* that humans are, nor of having the ability to pass on to their offspring what they have learned from their human mentors. Like the comparative psychologists of a previous era, these experimenters have failed to accept the probable fact that most of human representational activity is species-specific.

Nevertheless, the study of animal communication remains a fascinating area of investigation for biosemiotics. The question for the biosemiotician is not whether primates can speak like humans, but rather, what semiosic capacities they share with humans, and to what extent communicative interaction is possible between simians and people. It is likely that certain structural properties or features of semiosis cut across species, while others are specific to one or several species. Determining the universality or specificity of particular semiosic properties is a much more realizable goal, than is imparting the capacity for linguistic communication to nonhumans.

Biosemiotically, *communication* can be defined as *bilateral semiosis,* the capacity to participate with other organisms in the reception and processing of specific kinds of signals; *unilateral semiosis* is the capacity of an organism to receive and process specific kinds of signals in isolation. The systematic pattern of signal-exchanges in which organisms participate through bilateral semiosis defines the communication system for the species to which they belong. In the case of human communication, bilateral semiosis involves not only signaling behavior, but all kinds of representational form-exchanges. Engaging in bilateral semiosis with a member of another species entails sharing properties of the modeling system of that species. The greater the degree of commonality or parallelism of properties, the greater the likelihood that communication will be successful.

1.3 Singularized modeling

As mentioned above (§1.1), the most basic type of human representational form is the sign. The sign is a singularized form because it constitutes a simple model serving to encompass a singular referent or referential domain. In anthroposemiosis, nonverbal signs include gestures, bodily postures, facial expressions, tones of voice, visual forms (e.g., drawn figures); verbal signs include words, intonation patterns, graphic signs (alphabetic, ideographic, etc.).

In a biosemiotic paradigm, the function of singularized modeling is viewed as a general strategy for giving the perception of single objects, unitary events, individual feelings, etc. a knowable *form* (see also Thom 1975; Sebeok 1994). As von Uexküll (1973 [1928]: 40) argued, useful sensory information is something that a species would virtually not recognize, if it were not for the presence of an inbuilt modeling system designed to accomplish this task. Signs are, in effect, "recognition-enhancing forms", which allow for the detection of relevant incoming sensory information in a patterned fashion.

1.3.1 The sign

The sign has been defined in many different ways. In all definitions, however, three dimensions are implied: (1) a physical dimension (sounds, hand movements, etc.) termed the *signifier*(= [**A**]); (2) a conceptual dimension, termed the *signified* (= [**B**]), which elicits a singular referent or referential domain; (3) an interpretive dimension, termed *signification* (= [**A** ⊃ **B**]), which is the meaning (or meanings) extracted from a sign. The formula [**A** ⊃ **B**] is used here to highlight the fact that the two parts of the sign are inextricably interlinked: i.e., a sign is created the moment that a signifier is linked to a signified. Signifiers can, of course, exist without signifieds, as is evident in so-called nonsense words–i.e., well-formed words that do not have a meaning (*dop, flink, prip*, etc.). And, of course, there exists an infinitude of potential signifieds (referents or referential domains) that the current stock of signs in the world's languages, knowledge codes, etc. have not yet encoded.

1.3.2 Types of signs

Throughout the history of semiotics, there have been several attempts to identify and classify signs. Among these, Peirce's typology with 66 varieties, including intermediates and hybrids, is surely the most comprehensive, far-reaching, and sophisticated of all such attempts (Weiss and Burks 1945; Sanders 1970). In the verbal domain, one can also mention Roman Jakobson's (1970) classificatory system, which has shed considerable light on the minutiae of linguistic signification. Ignoring such minutiae for the sake of simplicity, six general categories of sign-making can be extrapolated from the relevant literature. These are: the *symptom*, the *signal*, the *icon*, the *index*, the *symbol*, and the *name*.

A *symptom* is a natural sign, recognizable by virtue of the fact that its signifier is coupled with its signified inside the body's morphology. It is a manifestation of some altered physical (histological, cytological, etc.) process, ranging from a painful sensation (such as headache or backache), to a visible condition (such as a swelling or a rash), or change in body temperature. A group of symptoms that collectively characterize a disease or disorder is called a *syndrome*. A syndrome is, therefore, a composite form with a fixed denotatum. Both terms have strong, but not exclusively, medical connotata. It is a peculiarity of symptoms that their denotata are generally different for the patient (subjective symptoms) than they are for the physician (objective symptoms) (Sebeok 1973b).

The bodies of all animals produce symptoms as warning signs, but what they indicate will depend on the species. As von Uexküll (1909) argued, the form of a symptom is a reflex of specific morphological structure. Animals with widely divergent bodily structures will manifest virtually no symptomatology in common. It is interesting to note, by the way, that in the human world of signification the term *symptom* is often extended metaphorically to refer to intellectual, emotional, and social phenomena that result from causes that are perceived to be analogous to physical processes: "Their behavior is a *symptom* of our times"; "Their dislike of each other is a *symptom* of circumstances"; etc.

The semiotician Roland Barthes (1972: 39) deemed it crucial to assign the symptom to the category of a pure signifier ([**A**]), turning into a sign ([**A** $\supset$ **B**]) only in the context of clinical discourse. However, such a view is tenable, if at all, only when the interpreter of a symptomatic form is a physician or, by extension, a veterinarian. In fact, the interpreter need be none of these; it could, for example, be a speechless creature (Darwin 1872: 101). Human symptoms can easily be perceived and acted upon by such domesticated animals as dogs and horses (Hediger 1967), in a variety of situations in which language plays no mediating role. In a biosemiotic perspective, then, the Barthesian conception of symptom is unwarranted.

The *signal* is a sign that naturally or conventionally (artificially) triggers some reaction on the part of a receiver (Sebeok 1972b: 514). Carpenter (1969: 44), a prominent researcher of animal behavior, defined signaling behavior as "a condensed stimulus event, a part of a longer whole, which may arouse extended actions". Like symptoms, signals are often excluded from consideration by some semioticians. Again, we find this position to be untenable.

All animals are endowed with the capacity to use and respond to species-specific signals for survival. Birds, for instance, are born with the instinctive capacity to produce a particular type of coo, and no amount of exposure to the songs of other species, or the absence of their own, has any effect on their cooing behavior. A bird reared in isolation, in fact, will sing a very simple outline of the sort of coo that would develop naturally in that bird born in the wild. This does not mean, however, that animal signaling is not subject to environmental or adaptational factors. Many bird species have also developed regional cooing "dialects" by apparently imitating each other. Vervet monkeys, too, have the usual set of signals to express emotional states and social needs, but they also have developed a particular predator signaling system–a specific call alerting the group to eagles, one to four-legged predators such as leopards, another to snakes, and one to other primates. The calls and general categories they represent seem innate, but in actual fact the young of the species learn them only by observing older monkeys and by trial and error. An infant vervet may at first deliver an aerial alarm to signal a vulture, a stork, or even a falling leaf, but eventually comes to ignore everything airborne except the eagle.

Most signals are emitted instinctively in response to particular types of stimuli and affective states. The response patterns suggest strongly that each animal species has a specific system of signals (Griffin 1981), although many are detected by other species. This is why human beings often read much more into animal behavior than is actually there. A well-known example of how easily people are duped by animal signaling is the case of *Clever Hans*. Clever Hans was heralded the world over as a German "talking horse" in 1904 who appeared to understand human language and communicate human-type answers to questions by tapping the alphabet with his front hoof–one tap for **A**, two taps for **B**, three taps for **C**, and so on. A panel of scientists ruled out deception and unintentional communication by the horse's trainer. The horse, it was claimed, could talk! Clever Hans was awarded honors and proclaimed to be an important scientific discovery. Eventually, however, an astute member of the scientific committee that had examined the horse, the Dutch psychologist Oskar Pfungst, discovered that Clever Hans would not tap his hoof without *observing* his questioner. The horse had obviously figured out–as most horses can–what the signals that his owner was unwittingly transmitting required him to do. The horse tapped his hoof only in response to inadvertent cues from his human handler, who would visibly relax when the horse had tapped the proper number of times. To show this, Pfungst simply blindfolded Clever Hans who, as a consequence, ceased to be so clever. The "Clever Hans phenomenon", as it has come to be known in the annals of psychology, has been demonstrated over and over with other animals as well (e.g., a dog will bark in lieu of the horse's taps in response to certain signals unwittingly emitted by people).

A large portion of communication among humans also unfolds largely in the form of unwitting bodily signaling. It has been shown, for example, that men are sexually attracted to women with large pupils, which convey unconsciously a strong and sexually tinged interest on the part of the female, as well as making the female look younger (Sebeok 1994). This would explain the fashion vogue in central Europe during the 1920s and 1930s of women using a crystalline alkaloid eye-drop liquid derived from the poisonous *belladonna* ('beautiful woman' in Italian) plant. The women of the day used this drug because they seemingly believed–and correctly so, it would appear–that it would enhance facial appearance and sexual attractiveness by dilating the pupils.

Humans are capable as well of deploying witting signals for some psychosocial purpose–e.g., nodding, winking, glancing, looking, nudging, kicking, head tilting. As the linguist Karl Bühler (1934: 28) aptly observed, such signals act like social regulators, eliciting or inhibiting some action or reaction. Artificial, mechanical, or electronic signaling systems have also been created for conventional social purposes. The list of such systems is extensive, and includes: smoke signals, semaphores, telegraph signals, warning lights, flares, alarms, sirens, bleepers, buzzers, knocking, bells, etc.

A sign is said to be *iconic* when the modeling process employed in its creation involves some form of simulation. Iconic modeling produces singularized forms that display a perceptible resemblance between the signifier and its signified. In other words, an *icon* is a sign that is made to resemble its referent in some way. Roman numerals such as I, II, and III are iconic signs because they imitate their referents in a visual way (one stroke = one unit, two strokes = two units, three strokes = three units); onomatopoeic words (*boom, zap, whack*, etc.) are also iconic signs because they constitute attempts to portray their referents in an acoustic way; commercially-produced perfumes that are suggestive of certain natural scents are likewise iconic because they attempt to simulate the scents in an artificial way; and the list could go on and on.

There are many manifestations of *iconicity* in zoosemiotic behavior as well, involving virtually all types of sensory channels–chemical, auditory, visual, etc. (Sebeok 1968: 614). Consider, for instance, the chemical signal that the ant species *Pogonomyrmex badius* emits. If the danger to the colony is momentary, some ants of this species release a chemical (warning) pheromone which quickly fades and leaves the bulk of the colony undisturbed; however, if the danger persists, the odorous signal spreads, animating an ever-increasing number of worker ants. The signal is iconic inasmuch as it varies in simulative proportion to the degree of danger faced by the colony (Sebeok 1982: 95).

The behavior of certain vespine audio-mimics, called *Spilomyia hamifera Lw*, is also classifiable as iconic. These flies display a wingbeat rate of 147 strokes per second while hovering near the wasp species *Dolichovespula arenaria F*–a species which the flies closely resemble in color pattern. Since the wasp flies with 150 wing strokes per second, the two flight sounds are indistinguishable to predators, and fly-catching birds are thus deceived (Sebeok 1972a: 86).

Finally, an elegant (if sometimes disputed) example of a complex form of signaling behavior that evolved, as it were, to function as a visual iconic form is graphically described by Kloft (1959). Kloft suggested that the hind end of an aphid's abdomen, and the kicking of its hind legs, constituted, for an ant worker, an iconic signifier, standing for the head of another ant together with its antennae movement. The ant can purportedly identify the likeness (the near end of the aphid) with its denotatum (the front end of an ant), and act on this information, i.e., treat the aphid in the manner of an *effigy*, which is a visual icon.

A sign is said to be *indexical* when its representational focus is the location of a referent in space, time, or in relation to some other referent. In one of his most memorable examples, Peirce referred to the footprint that Robinson Crusoe–the character created by British novelist Daniel Defoe (1660-1731) in his 1719 novel of the same name–found in the sand, which was interpreted by Crusoe as an index of some creature. In actual fact, a vast map of such indexical marks is printed overnight by animals of all sorts (Ennion and Tinbergen 1967: 5). Indexes do not resemble their referents, like icons do; rather, they indicate or show where they are in relational terms. The most typical manifestation of *indexicality* is the pointing index finger, which humans the world over use instinctively to point out and locate things, people, and events in the world. Many words, too, manifest indexicality: e.g., *here*, *there*, *up*, *down*, etc.

Indexicality is more technically a manifestation of *deixis*, the process of referring to something by pointing it out or specifying it in some way (from Greek *deiktos* 'able to show directly'). There are three types of deixis:

- *Spatial deixis* is the process of referring to the spatial locations of objects, beings, events, etc. Examples of spatial deictic forms are the pointing index finger, demonstratives such as *this* or *that*, adverbs such as *here* or *there*, etc.
- *Temporal deixis* is the process of referring to the temporal relations that exist among things and events. Examples of temporal deictic forms include adverbs such as *before*, *after*, *now*, or *then*, timeline graphs representing points in time, ordinal numerals (*first, second,* etc.), and so on.
- *Personal deixis* is the process of referring to the relations that exist among participants taking part in a situation. Personal deictic forms include personal pronouns like *I, you, he, she*, indefinite pronouns like *the one, the other*, etc.

A singularized form is *symbolic* when the modeling process employed in its creation is constrained by cultural and historical factors. Most semioticians agree that *symbolicity* is what sets human modeling apart from that of all other species, allowing human beings to represent things independently of stimulus-response situations. Many words are used symbolically. But any signifier–object, sound, figure, etc.–can be used symbolically: e.g., a cross figure can stand for the concept [Christianity]; a V-sign made with the index and middle fingers can stand for the concept [peace]; the color *white* can stand for [cleanliness], [purity], [innocence], and *dark* for [uncleanness], [impurity], [corruption]; and the list could go on and on. The ability to model the world symbolically is evidence that human consciousness is not only attentive to sensible properties (resulting in iconic modeling activities), and to spatiotemporal and relational patterns (resulting in indexical modeling activities), but also to all kinds of referents (actual and potential) in and of themselves.

A *name* is a form that identifies a human being (*Alexander, Sarah,* etc.) or, by connotative extension, an animal, an object (such as a commercial product), or event (such as a hurricane). A *name* has both indexical and symbolic properties: it is partly an indexical form because it identifies a person and, usually, points to h/er ethnic origin; it is partly a symbolic form because, like any word, it is a product of conventionalized representational practices. Less often, names are coined iconically: Trivial but instructive examples of this can be seen in the names given typically to household animals–*Ruff, Pooh-Pooh,* etc.

1.3.3 Sign-making principles

The principles enunciated above (§1.2.1) are intended to explicate various sign-making phenomena in an integrative fashion–phenomena that would otherwise be treated as unrelated. Take, for instance, the dimensionality principle. This posits that the making of iconic and indexical forms of representation is prior to the making of symbolic ones. The developmental evidence supporting this principle is substantial. When the child becomes aware of objects, such as toys, s/he instinctively points to them (indexicality) and/or imitates any sounds they make (iconicity) while playing with them; only at a later stage does s/he start referring to objects by name (symbolicity). As a corollary, the dimensionality principle identifies iconicity and indexicality as the default modes of representation. The validity of this principle becomes obvious every time people instinctively resort to gesturing, making imitative sounds, and pointing to things when communicating with those who do not speak the same language they do.

The same principle also implies that many singularized forms are amalgams of iconic, indexical, and symbolic modeling strategies. A trivial, but nevertheless instructive, manifestation of dimensionality is the way in which the common traffic sign for a crossroads is constructed:

Figure 5. Crossroads traffic sign

The signifier of this sign consists of two straight lines intersecting at right angles. This cross figure this makes is both iconic and symbolic: it is iconic because its shape visually resembles a [crossroads]; but since the cross figure could easily be used to represent a [church] or a [hospital] in other situations (without the arrowhead), it is also symbolic insofar as is conventionalized use on a road sign refers to a [crossroads]. Finally, the sign is indexical because when it is placed near an actual crossroads it indicates that one is about to reach it physically.

1.4 Composite modeling

Composite modeling is the activity of representing complex (non-unitary) referents by combining various signifiers in some specifiable way. Drawings, narratives, theories, conversations, etc. are all examples of composite forms of representation. These are constructed with distinct signifiers that fit together structurally, but which are, as a whole, different from any of their constituent signifiers taken individually. In analogy to atomic theory, a singularized form can be compared to an atom and a composite form to a molecule made up of individual atoms, but constituting a physical form in its own right.

1.4.1 The text

Texts incorporate the structural properties of the signifiers with which they are constructed, but they are not conceptually equivalent to the aggregate of their signifieds. A novel, for instance, is made up of words following one after the other. But conceptually it is not just the sum of the meanings of the words; rather, a novel constitutes a composite form that generates its own signified(s). This is why people do not interpret a novel in terms of its constituent parts, but holistically as if it were single sign, [**A** ⊃ **B**], where [**A**] is the novel and [**B**] the meaning(s) people extract from it: e.g., "The novel *Crime and Punishment* ([**A**]) paints a grim portrait of the human psyche ([**B**])".

Drawings, theories, and other composite forms are interpreted in this fashion. For instance, when asked what the theory of relativity is all about, people will typically couch their answer as follows: "The theory of relativity ([**A**]) explains how time and space are interrelated ([**B**])". One can, of course, relate the signifying parts to each other in an interpretive discussion of the text. This is, in fact, what people do when they discuss a novel's meaning by referring to parts of the novel, to its plot, to its characters, etc. But in all such discussions, the parts are related to the signified ([**B**]) extracted from it, rather than seen as separate from it.

The meaning of a text is conditioned by *context*. The context is the situation–physical, psychological, and social–in which a text is constructed, used, occurs, or to which it refers. Consider a discarded cigarette butt. If one were to come across this object on a sidewalk on a city street, one would no doubt view it as a piece of rubbish. But if one saw the very same object displayed in an art gallery, "signed" by some artist, and given a title such as "Waste", then one would interpret it as an *artistic text*, and thus assign a vastly different meaning to it. Clearly, the cigarette's physical context of occurrence and social frame of reference–its location on a sidewalk vs. its display in an art gallery–will determine how one will interpret it. Human beings are capable of constructing an infinite number of novel texts appropriate to a limitless number of contexts.

1.4.2 Types of texts

There are as many types of composite forms as there are singularized ones. For example, syndromes are, in effect, *composite symptoms* which collectively indicate or characterize a disease, a psychological disorder, or some other abnormal condition. *Composite signaling* can be seen, for instance, in courtship behaviors where various sensory signals (visual, auditory, olfactory, tactile, gustatory) often coalesce to stimulate mating urges. An example of an *iconic composite form* is an imitative drawing of a scene. An *indexical composite form*, such as a typical map, is one that is constructed to refer to spatial or temporal phenomena in an integrative relational way. A *symbolic composite form*, such as a mathematical theory, is a text that is made with the symbolic resources of a culture. Finally, a *composite name* consists of several identifiers (e.g., [given name] + [surname]) providing various kinds of culture-specific information–e.g., where the person is from, what h/er parentage is, etc. Composite modeling occurs in all facets of human life, allowing people to envision distinct bits of information and real-world phenomena as integrated wholes. As the mathematician Sawyer (1959: 34) has suggested, these models probably manifest the details of how the mind abstracts knowledge from the world.

But composite modeling is not a specific capacity of human semiosis. It is, in fact, found in other species. One well-known example is the honeybee dance. Worker honeybees returning to the hive from foraging trips inform the other bees in the hive about the direction, distance, and quality of the food with amazing accuracy through movement sequences which biologists call a "dance", in obvious analogy to human dancing. The remarkable thing about the dance is that it appears to share with human representation the feature of displacement, i.e., of conveying information in the absence of the referential domain to which it calls attention.

Several kinds of dance patterns have been documented by entomologists. In the "round" dance, the bee moves in circles alternately to the left and to the right. This dance is apparently deployed when the cache of food is nearby. When the food source is further away, then the bee dances in a "wagging" fashion, moving in a straight line while wagging its abdomen from side to side and then returning to its starting point. The straight line in this dance points in the direction of the food source, the energy level indicates how rich the food source is, and the tempo provides information about its distance. In one experimental study a feeding dish placed 330 meters from the hive triggered a dance which consisted of 15 complete rounds in 30 seconds, whereas a dish located 700 meters away triggered a dance which consisted of only 11 runs carried out in the same period of time (Frisch (1962).

1.4.3 Text-making principles

The making of composite forms is guided by the same principles that undergird the construction of singularized ones (§1.2.1). Dimensionality, for instance, can be discerned in the make-up of maps. Although the signifiers of a map are ultimately interpreted symbolically, since their representational functions are products of conventional practices and must therefore be learned in cultural context, the ways in which they are put together reveal iconic, indexical, and analogical properties: e.g., the scale to which a map is drawn represents the ratio of the distance between two points on the earth and the distance between the two corresponding points on the map (both an indexical and an analogical property); the varying heights of hills and mountains, and the depths of valleys and gorges as they appear on a topographic map, show in miniature how these relate to each other in real space (an iconic property); the shapes of contour lines provide an accurate representation of the shapes of hills and depressions (an iconic property); and so on.

The interconnectedness principle manifests itself, for instance, in what is known in semiotic theory as *intertextuality* (e.g., Bernardelli 1997), i.e., in the fact that a specific text bears meaning in a culture because it often alludes (in part or in whole) to already existing texts. For example, allusions to religious themes abound in novels, making their decipherment dependent upon knowledge of the culture's religious themes and images. Extracting a meaning from John Bunyan's (1628-1688) novel *Pilgrim's Progress* (two parts, 1678 and 1684), for instance, is contingent upon knowing the Bible narrative, since it constitutes an allegorical tale of a Christian's journey from the City of Destruction to the Celestial City.

Interconnectedness also manifests itself typically in the fact that a specific type of composite model is often constructed (in part or in whole) with the modeling resources of a code that was devised for some different purpose. For example, maps in Western culture are typically constructed with the principles of the code of Cartesian coordinate geometry, which was invented for a specific mathematical, not topographic, purpose (to unify algebra and geometry). The use of latitude and longitude lines to define locations on a map involves knowledge of this cohesive modeling system, in which a point in the plane is defined in terms of coordinates which relate its location with respect to two perpendicular lines that intersect at an origin.

1.5 Cohesive modeling

A cohesive modeling system is known in traditional semiotic theory as a code (§1.1), a system providing particular types of signifiers that can be used in various ways and for diverse representational purposes. A cohesive modeling system can be compared to a computer program or to a common recipe. The former consists of a set of instructions that the computer can recognize and execute converting information from one form into another; the latter of a set of directions for preparing something to eat or drink by combining various ingredients. A language code, for instance, provides a set of phonetic, grammatical, and lexical "instructions" that the producers and interpreters of words and verbal texts can recognize and convert into messages.

Generally speaking, for some particular representational need there is an optimum code or set of codes that can be deployed. For example, the composer of a work of operatic art will need to deploy at least three code-making sources in the construction of h/er text: the musical code, the verbal code, and the theater code (all in place at the time of the composition).

1.5.1 The code

The system of Cartesian coordinate geometry (§1.4.3) is a perfect example of what constitutes a code. In two dimensions, this is a cohesive modeling system based on dividing the plane into four quadrants with two axes crossing at right angles, the *x*- and *y*-axes. The point at which they intersect is called the origin.

The point (2, -3), for instance, is definable by its location with respect to these two axes: i.e., it is located two equally-calibrated units to the right of the *y*-axis and three units down from the *x*-axis (figure 6); similarly, the point (-3, -1) is located three units to the left of the *y*-axis and one unit down from the *x*-axis, and the point (3, 1) three units to the right of the *y*-axis and one unit up from the *x*-axis.

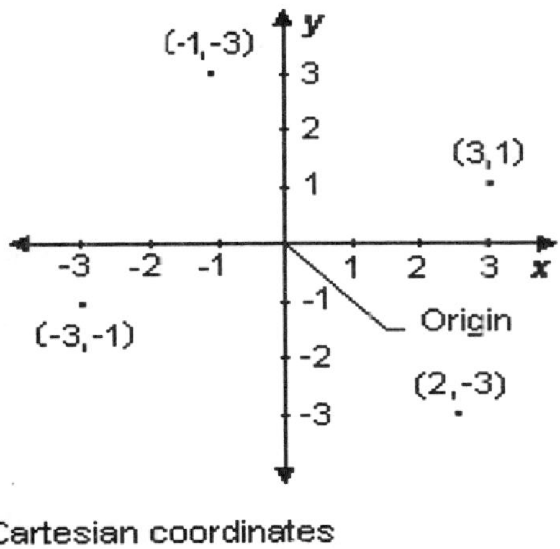

Figure 6. Points in the Cartesian plane

In this two-axial system, points are identifiable in relation to each other in terms of such coordinates. Moreover, points can be joined to produce specific kinds of geometric figures. For instance, by joining (2, -3) and (3, 1), we get the figure of a *line* (figure 7); and by joining the points (2, -3), (3, 1), and (-3, -1), we get the figure of a *triangle* (figure 8).

This system makes it possible to model geometrical concepts in specific ways. This code is, therefore, an example of a cohesive modeling system of two-dimensional space. It can be extended to represent points in three-dimensional space, with the use of three axes (*x, y, z*) representing the distances from three planes determined by three intersecting straight lines not all in the same plane; i.e., with the *x*-coordinate representing the distance from the *yz*-plane measured along a parallel to the *x*-axis, the *y*-coordinate representing the distance from the *xz*-plane measured along a parallel to the *y*-axis, and the *z*-coordinate representing the distance from the *xy*-plane measured along a parallel to the *z*-axis (the axes are usually taken to be mutually perpendicular). Analogous systems may be constructed for describing points in abstract spaces of four or more dimensions.

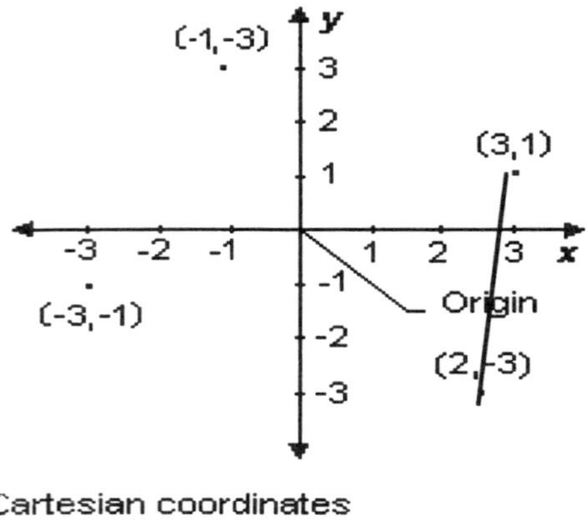

Figure 7. Deriving the figure of the line in the Cartesian plane

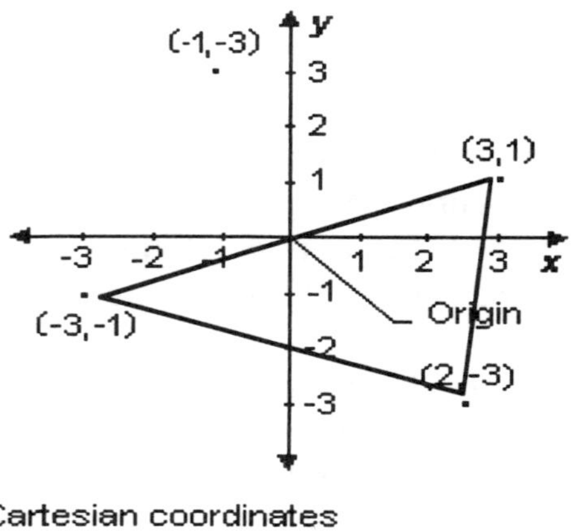

Figure 8. Deriving the figure of the triangle in the Cartesian plane

The use of a code to make signs or texts is called *encoding*; the reception and interpretation of signs or texts is called *decoding*. The Canadian communications theorist Marshall McLuhan (1964) pointed out that encoding and decoding processes are shaped by *sense ratios*, as he called them, which are equally calibrated at birth. In specific social settings, it is unlikely that all senses will operate equally in the encoding/decoding process. One sense ratio or the other increases according to the types of codes and media employed. In an oral culture, the *auditory sense ratio* dominates encoding and decoding processes; in an alphabetic one, the *visual sense ratio* dominates instead. This raising or lowering of a sense ratio is not preclusive. Indeed, in modern cultures, it is typical to utilize several sense ratios in tandem. The ebb of ratios, up and down, in tandem, in opposition, is what defines one's *cognitive style* of information processing.

As a concrete example, consider the word *ball*. If one were to hear this uttered vocally, then h/er auditory sense ratio would be the more operational one in decoding the meaning of the word. If, however, one were to see the word written on a sheet of paper, then h/er visual sense ratio would be the operational one. A visual depiction of the ball together with the utterance of the word *ball* on TV (as in children's learning programs) would activate the auditory and visual sense ratios in tandem. In effect, the way a sign or text is processed cognitively depends on the code deployed and the medium through which it is transmitted.

Media of transmission and codes are intertwined in encoding and decoding. If the sign or text is transmitted through an auditory medium, then the phonemic code of a language must be known, otherwise encoding/decoding is impossible; if it is written on a piece of paper, then the alphabetic code of the language must be known; and so on. In effect, the medium determines which code is to be deployed in making forms, and vice versa. This is probably what McLuhan meant when he said the "medium is the message" (figure 9).

The same sign or text can, of course, be encoded in more than one way–in an oral medium (e.g., an oral story), an alphabetic medium (e.g., a novel), a multisensory medium (e.g., a movie), etc. It will be thus decoded according to the characteristics of the medium (or media) deployed. So, for instance, the story of Romeo and Juliet can be transmitted to someone orally, activating the auditory sense ratio; it can be conveyed through the novel form, activating the visual sense ratio; it can be portrayed through cinema, activating several sensory ratios in tandem.

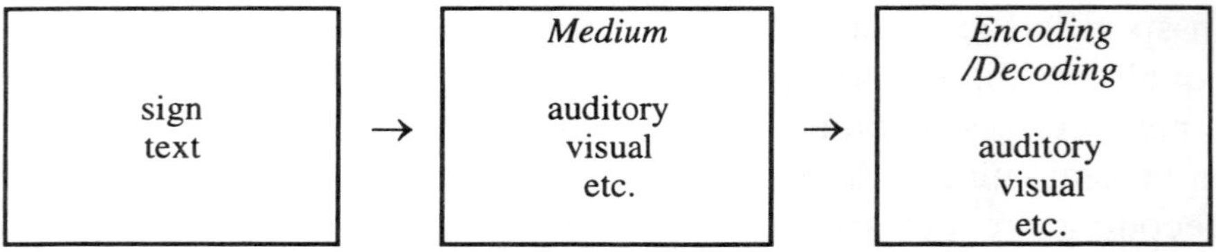

Figure 9. Correspondence of encoding/decoding system used to medium deployed in sign and text creation

1.5.2 Types of codes

There are as many types of codes as there are signs or texts. For example, the body's immune system is a natural code consisting of interacting organs, tissues, cells, and cell products such as antibodies which not only neutralize potentially pathogenic organisms or substances, but also allow one to become aware of the difference between Self and "nonSelf" (the external world). It is the code that undergirds the symptomatology of diseases. An example of a simple mechanical (artificial) signaling code is the common traffic light system: a red light, green light, or yellow light inform a driver or pedestrian to stop, move forward, or slow down respectively. The Roman numeral system is an example of an artificial code fashioned in part iconically. This system consists of seven symbols for representing all numbers from 1 to 1,000,000: I for 1, V for 5, X for 10, L for 50, C for 100, D for 500, and M for 1000. The main iconic feature of this code is that one stroke represents one unit, two strokes, two units, three strokes three units: I = [one unit], II = [two units], III = [three units]. An example of an indexical code is the system of street signs used typically in modern industrialized societies. These signs provide information, among other things, about the distance of certain places from specific locations, about the direction one is traveling in, etc. An example of a simple symbolic code is the *Morse code* (which is now no longer in use). This allowed people in the not-too-distant past to make verbal texts with dots and dashes (a dash is equal to three dots in duration) which were transmitted by a flash lamp, telegraph key, or other device. A letter or a number was represented (conventionally) by a combination of dashes and dots.

1.5.3 Code-making principles

Cohesive modeling systems are governed by the same general principles that govern singularized and cohesive modeling strategies. The extensionality principle, for instance, manifests its presence in the number-making code known as the Arabic numeral system. Although the signifiers of this code (numerals), and the numerical texts made with it (actual numbers), can only be interpreted symbolically, since their representational properties are products of conventional practices and must therefore be learned in cultural context, the origin of some of these reveal iconic properties: the numeral 1, for instance, is analogous to the Roman numeral I in that it constitutes one stroke representing one unit; the numeral 3 too is, in effect, a stylized and rotated version of three strokes representing three units.

Codes are also interconnected to each other in a culture. This manifestation of interconnectedness can thus be called *intercodality*. For instance, understanding language texts involves knowledge of several other codes: e.g., the phonemic code (and the alphabetic code if the text is written); the discourse code (how a text is to be delivered in social settings), etc. Language codes are also highly interconnected with gesture codes. When people are talking they typically use gestures to reinforce, exemplify, or even elaborate upon what they are saying.

1.6 Connective modeling

Connective forms are the result of metaphorical reasoning processes (recent summaries of relevant work on metaphor can be found in Gibbs 1994 and Goatley 1997). In our view, the ever-burgeoning literature on what has come to be known as *conceptual metaphor theory* (e.g., Lakoff and Johnson 1980, 1999; Lakoff 1987; Johnson 1987) is highly intriguing, but still lacks a synthetic semiotic framework for interpreting the diverse, multiform manifestations of metaphor in human symbolic and communicative behavior. This framework is, also in our view, provided by MST and, more specifically, by the derived notion of *metaform.*

1.6.1 The metaform

A *metaform* is an example of a *connective form* that results when abstract concepts are represented in terms of concrete ones. The formula [thinking = seeing], for example, is a metaform because it delivers the abstract concept of [thinking] in terms of the signifieds associated with the concrete concept of [seeing]. This metaform (which is a largely unconscious mental form in native speakers) underlies utterances such as:

1. I do not *see* what possible use your *ideas* might have.
2. I can't quite *visualize* what that new *idea* is all about.
3. Just *look at* her new *theory*; it is really something!
4. I *view* that *idea* differently from you.

Each of the two parts of the metaform is called a *domain*: [thinking] is called the *target domain* because it is the abstract topic itself (the "target" of the representation); and [seeing] is called the *source domain* because it enfolds the class of vehicles (forms with concrete signifieds) that deliver the meaning of the metaform (the "source" of the metaphorical concept) (Lakoff and Johnson 1980). A specific metaphorical statement uttered in a discourse situation is now construable as a particular externalization of a metaform. So, when we hear people using such statements as those cited above, it is obvious that they are not manifestations of isolated, self-contained metaphorical creations, but rather, specific instantiations of the metaform whose target domain is [thinking] and whose source domain is identifiable as [seeing]:

Metaform = [thinking = seeing]

↓

I don't *see* ...I can't quite *visualize* ...Just *look at*...I *view*...

↑

Specific externalizations (Actual metaphorical statements)

Figure 10. Difference between a metaform and a metaphor

Psychologically, metaforms relate the experience or understanding of some abstract notion to something that is familiar and easily perceivable in both imagistic and representational terms. In more strict semiotic terms, they reveal a basic tendency of the human mind to think of abstract concepts iconically. This aspect of human thinking was, needless to say, noticed originally by Aristotle (384-322 BC), who, as is well known, coined the term *metaphor*, because he noticed how figurative speech was used consistently to make abstract concepts "knowable" in concrete terms. The sapient animal is a metaphorical thinker, Aristotle contended. However, Aristotle also affirmed that, as knowledge-productive as it was, metaphor's most common function was to spruce up literal ways of thinking and speaking. Remarkably, it was this latter position of Aristotle that became the accepted view of metaphor in Western society at large, until recently.

1.6.2 Types of connective forms

Metaforms have become the target of great interest in linguistics and psychology in the last five decades (e.g., Gibbs 1994; Fisher 1998; Lakoff and Johnson 1999). The gist of the research points to metaphor and metonymy as the primary processes involved in connective modeling, revealing that most abstractions are, in effect, "informed best guesses" based on concrete experiences. There are three main types of connective models: *metaforms*, *meta-metaforms*, and *metasymbols*.

The difference between a *metaform* and a *metaphor* is, in effect, one of hyponymy. As discussed above (§1.6.1), a specific metaphor is a verbal instantiation of a metaform. Metaforms are *primary connective forms*, portraying abstractions in terms of concrete source domains. The [thinking = seeing] metaform, for instance, is linked to how we conceptualize [ideas], [theories], [awareness], [discernment], [clarification], [perspective], etc. (Danesi 1990). These abstract notions are all conceived as *ways of seeing internally* that are modeled on *ways of seeing externally*.

Now, once the first "layer" of metaforms has been formed in a language's conceptual reservoir, on the basis of concrete source domains, then this layer itself becomes a new productive source domain for creating a higher (= more abstract) layer of concepts. This has been called the *layering principle* elsewhere (Danesi 1999a). The forms resulting from linkages among metaforms can be called *meta-metaforms*. Thus, for example, in utterances such as the following the target domain of [thinking] is rendered by source domains that are themselves metaforms: namely, [thinking = upward motion] and [thinking = scanning motion].

5. When did you *think up* that idea?
6. I *thought over* carefully your ideas.
7. You should *think over* the whole problem before attempting to solve it.

Phrasal verbs such as *think up* and *think over* are, in effect, products of an association of [thinking] with an [upward motion] and with a [scanning motion] respectively. A linkage of these two produces the meta-metaform: [thinking = upward + scanning motions] as in the following:

8. That idea *came out* of *nowhere*.
9. That theory *emerged* from the *landscape* of my thoughts.

Expressions such as *come out of nowhere* and *emerge from the landscape* are products of the meta-metaform [thinking = upward + scanning motions].

The third kind of connective form is really a type of symbol. For example, a *rose* is used as a symbol for *love* in Western culture because its physical features–*rose* = [sweet smell], [red color], [plant]–also constitute source domains for [love]: namely, [love = sweet smell], [love = red color], and [love = plant]. This is how the symbol [rose = love] came about. It is an example, therefore, of a *tertiary connective model*, which can be called, more specifically, a *meta-symbol*.

In summary, the three main types of connective models are: (1) *metaforms*, which are assemblages intended to deliver the meaning of abstract concepts on the basis of concrete source domains; (2) *meta-metaforms*, which are assemblages forged among already-existing metaforms; (3) *meta-symbols*, which are symbolic forms that result from specific types of linkages associated with particular metaforms:

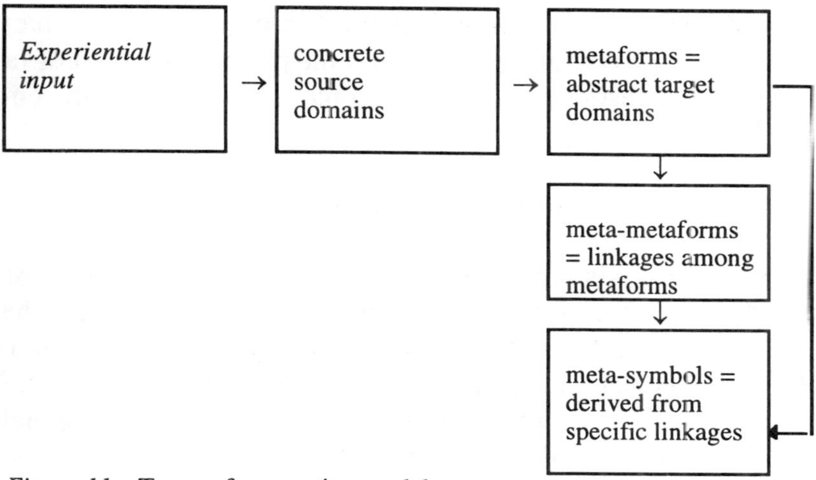

Figure 11. Types of connective models

Before the trend-setting work on metaphor within linguistics, the study of figurative speech fell within the field of *rhetoric*, where it was viewed as one of various *tropes*–i.e., figures of speech. But since the early 1980s the practice has been to use the term *metaphor* to encompass virtually all forms of connective modeling. Personification, for instance, (e.g., "My cat speaks Hungarian") can be seen as a particular kind of metaform, one in which the target domain is an animal and the source domain some human characteristic: [animals have human characteristics].

A metaform may also be the product of metonymic reasoning. *Metonymy* entails the use of an entity to refer to another that is related to it. A metonymic metaform results when part of a domain starts being used to represent the whole domain (Lakoff and Johnson 1980: 35-40):

10. She likes to read Dostoyevski (= the writings of Dostoyevski).
11. He's in dance (= the dancing profession).
12. My mom frowns on blue jeans (= the wearing of blue jeans).

13. Only new wheels will satisfy him (= car).

Each one of these constitutes an externalization of a metonymically-derived metaforms: (8) is an instantiation of [the author = h/er work], (9) of [an activity of a profession = the profession], (10) of [a clothing item = a lifestyle], and (11) of [a part of an object = the entire object].

1.6.3 The species-specificity of connective modeling

Connective modeling is unique to anthroposemiosis. The equivalent of metaforms, meta-metaforms, and meta-symbols are nowhere to be found in the zoosemiotic world. Phylogenetically speaking, the universality of connective modeling in the human species begs the question of the relation of metaphor to the emergence of conceptual thinking in humans. Indeed, the crystallization of primary connective forms in human representation suggests that sensory perception was originally at the root of many abstract notions. Thus, for instance, the source of the [love = a sweet taste] metaform in English–"She's my *sweetheart*"; "I love my *honeybunch*"; etc.–appears to be the pleasant gustatory experience that the romantic kiss of a loved one, for instance, produces. So, the linkage of [a sweet taste] to [love] is hardly a random or fanciful process. It is based, rather, on having experienced the latter in some real-world way. In this book, we will refer to this type of experientially-based conceptualizing as the *Sense-Inference Hypothesis* (§2.1.2).

Once metaforms such as [thinking = seeing] or [love = a sweet taste] have entered the language and the signifying order, they become, as discussed above, new source domains for further linkages. We will discuss this phenomenon in subsequent chapters. It is sufficient to say here that such linkages characterize complex human mental activity. The vast empirical literature on metaphor now suggests that most attempts at abstract representation are ultimately based on complex connective assemblages of various sorts.

In a fundamental semiotic sense *culture* can be defined as a *connective macrocode*, made up of the different codes (language, gesture, music, etc.) and the signs, texts, and connective forms that are fashioned and used by people in specific social contexts. This macrocode constitutes a *signifying order*, which can be defined as a an interconnected system of signs, texts, codes, and connective forms:

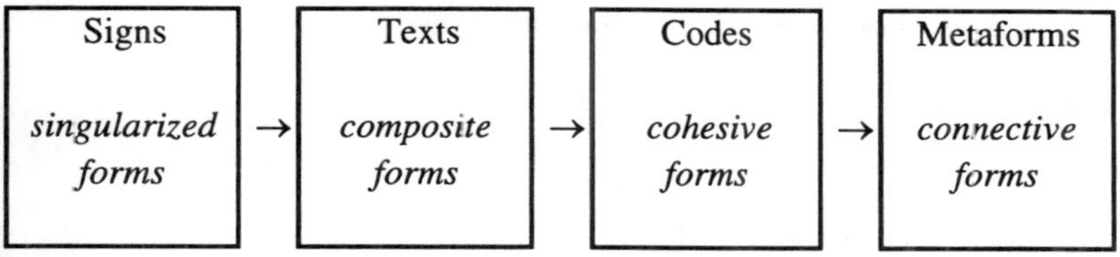

Figure 12. The signifying order

One of the more fundamental questions that this line of investigation begs is: Are all abstractions and symbols based on metaphorical reasoning? As Levin (1988: 10) has aptly remarked, there appear to be many kinds of concepts and modes of knowing: "innate knowledge, personal knowledge, tacit knowledge, spiritual knowledge, declarative and procedural knowledge, knowing that and knowing how, certitude (as well as certainty), and many other varieties". The more appropriate goal for future research on connective representation within an MST framework should be, therefore, to determine to what extent connective forms are used to encode abstractions and what other types are available for this same purpose.

Chapter II
Primary modeling

Nature is commonplace. Imitation is more interesting.

Gertrude Stein (1874–1946)

2. Introductory remarks

The starting point of *Systems Analysis* (SA) is a detailed study of the manifestations of primary modeling phenomena across species. As defined in the opening chapter (§1.2), the PMS is the innate capacity for simulative modeling, i.e., it is the system that underlies forms produced by the simulation of some sensory property of a referent or referential domain. In the human semiosic realm, the PMS manifests itself in singularized, composite, cohesive, and connective modeling phenomena: e.g., joining the thumb and index fingers to represent a circular object constitutes produces a singularized form; reproducing scenes in a painting produces a composite model; simulating certain bodily features (as in erotic dancing) constitutes cohesive modeling; using the [love = a sweet taste] metaform in discourse situations (§1.6.3) constitutes connective modeling.

The two chief objectives of SA are: (1) to document and catalogue the manifestations of primary modeling forms across species; and (2) to look at how the human PMS allows people to encode their sensory impressions of the world through the creation of a limitless number of singularized (iconic signs), composite (iconic texts), cohesive (iconic codes), and connective (metaforms) representational forms. The inventory of such manifestations can then be examined in the light of the biosemiotic principles discussed in the previous chapter (§1.2.1).

2.1 The primary modeling system

The notion of primary modeling was developed primarily by the Moscow-Tartu School of semiotics in the early 1960s (Lucid 1977:

47-58; Rudy 1986). As subsequent biosemiotic research on this notion has shown, there exist two distinct kinds of primary modeling processes. Although they have been assigned various names in the relevant literature, in this book they will be designated *osmosis* and *mimesis*. Osmosis refers to the spontaneous production of a simulative form in response to some stimulus or need; mimesis refers instead to the intentional making of forms in a simulative manner. In a phrase, osmosis is natural (unwitting) simulation, mimesis intentional (witting) simulation. The nodding gestures a speaker makes in synchrony with an interlocutor as s/he speaks, conveying agreement, empathy, etc. with what the interlocutor is communicating vocally, are examples of osmotic (spontaneous) forms. On the other hand, the facial expressions a stage actor makes on purpose to convey some emotional state to an audience are examples of mimetic (intentional) forms.

In general, the PMS can be characterized as the modeling system that allows organisms to simulate something in species-specific ways. In the three-system (PMS, SMS, TMS) anthroposemiotic domain, the PMS is the "default" system. This is evidenced by the fact that initial attempts to encode a referent typically reveal a recourse to, or reliance upon, primary modeling (Lotman and Uspenskij 1978). Needless to say, once a form has been created in a simulative way, and then gains currency in a social context, it can be extended to encompass abstract referents. For example, the English word *boom* was obviously coined initially as a primary singularized model in phonic simulation of a [sudden deep resonant sound], as in: "That explosion went *boom*"; I heard a *booming* sound just behind the bushes"; Did you hear the *boom* in my car engine"? etc. Now, this very same signifier can be applied to abstract referents or referential domains (connotata) that are felt, by extension, to involve [suddenness]. One such connotatum is: [sudden spurt of growth]– e.g., "Their business is *booming*"; "They are part of the *baby-boom* generation"; "It's *boom* or bust for the economy this year"; The *boom* in housing starts this year is due to increased affluence"; etc. Connotata such as this one are products of *extensional modeling* (§1.1.3), which is a secondary modeling process, as will be discussed in detail in the next chapter. The above signification process, or "modeling history", as it can be called, is shown below in figure 13.

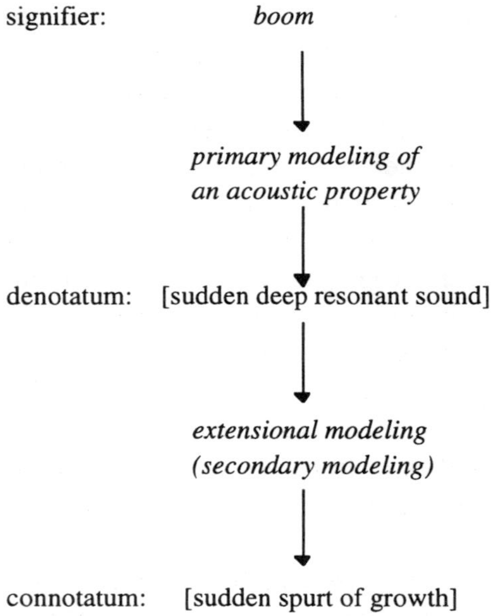

Figure 13. Modeling history of *boom*

2.1.1 Natural and intentional simulation

Osmotically-produced forms are found across species. Differences can be traced, as von Uexküll (1909) insisted, to the distinct kinds of anatomies and sensory apparatus possessed by diverse species (Gipper 1963; Sebeok 1990, 1991a). As Jacob (1982: 55) has aptly put it, every species "is so equipped as to obtain a certain perception of the outer world, thus living in its own unique sensory world, to which other species may be partially or totally blind". In effect, osmotic simulation is an inbuilt capacity of all species that varies according to the particular physical characteristics possessed by the species.

Take, as an example, the phenomenon of camouflage. This is definable as the osmotic process whereby some aspect of a species' physical appearance undergoes changes that make it seem to be part of its surroundings. The adult females of the species known as scale insects (*Icerya purchasi* and *Quadraspidiotus perniciosus*), for instance, attach themselves by their mouthparts to plant and tree surfaces, secreting a waxy substance that makes them appear to be part of those surfaces. The common leaf insect (*Phyllidae*) has the capacity to generate enlargements of its legs and abdomen, which make it resemble leaves. Similarly, any of several species of long-horned grasshoppers called katydids (*Tettigoniidae*) have developed broad wings that resemble leaves in appearance, allowing them to blend in with their environment.

In the human semiosic realm, osmosis manifests itself typically in the spontaneous formation of facial expressions to convey emotional states. Eyebrow position, eye shape, mouth shape, and nostril size constitute osmotic signifiers that reveal such states. When someone is telling a lie, for instance, the pupils tend to contract, one eyebrow may lift, and the corner of the mouth may twitch.

Mimetic modeling is not mere copying, but rather, a simulative process guided, in large part, by the conscious inferences people make about the perceptible features of the referents they are attempting to encode. We will refer to this as the *Sense-Inference Hypothesis* (SIH). The word *boom* discussed above (§2.1), for example, is a product of mimetic modeling. In all likelihood, the person who coined this word, did so in order to imitate a [sudden deep resonant sound] feature of some referent with a subjective deployment of the phonemic resources of English. H/er creation of this form was, in effect, guided by h/er "sense inference" of a certain sound constrained only by the English sound system and h/er specific utilization of that system. In actual fact, this word derives from Old German *bummen*, and the English word is a phonetic adaptation of that etymon.

The SIH traces its antecedents to the ideas of such philosophers as John Locke (1632–1704)), Giambattista Vico (1688-1744), Ernst Cassirer (1874-1945), and Suzanne Langer (1895-1985), among others. The English philosopher Locke argued that all forms of representation were based initially on the categories of sensation, with reflection being no more than the internal state of consciously *recognizing* what the senses have previously *cognized.* The Italian philosopher Vico saw symbolic thinking as the end-product of sensory knowing. The mind, Vico emphasized, "does not understand anything of which it has had no previous impression from the senses" (in Bergin and Fisch 1984: 123), because it is "naturally inclined by the senses to see itself externally in the body; and only with great difficulty does it come to understand itself by means of reflection" (in Bergin and Fisch 1984: 95). The German philosopher Cassirer viewed concept-formation as, essentially, sense-extension. The American philosopher Langer saw all efforts on the part of human beings to understand inner and outer experiences as basically sensory in origin. The "feeling structure" inherent in these experiences, Langer claimed, is subsequently converted into "analytical structure" in the domain of language.

2.1.2 Iconicity

In the human semiosic domain, simulative forms–known more specifically as icons–abound, spanning the entire range of sense inference:

- Onomatopoeic words such as *drip*, *plop*, *bang*, *screech* are vocal singularized icons that have been made to replicate the sounds that certain actions or movements are perceived to make.
- Portraits of people are visual composite icons that have been made to replicate the actual faces of people from the perspective of the portrait-maker.
- A perfume fragrance is an olfactory singularized icon that has been made to simulate some natural scent.
- Chemical food additives are gustatory singularized icons that have been manufactured to simulate the tastes of natural foods.
- A block with a letter of the alphabet carved in relief is a tactile singularized icon that has been made to reproduce the letter in terms of its actual shape.

The SIH rejects the Saussurean view of human semiosis as essentially constrained by conventionalized, basically arbitrary modeling processes. But, as it turns out, even the most abstract form produced by human effort will reveal, upon close scrutiny, an iconic origin. As an example, consider the word *flow*. The denotatum of this word encodes the action of moving or running smoothly with unbroken continuity, a property that can be observed to be characteristic of a fluid. This property can be represented, for the sake of convenience, as the feature [fluidity]. A strict Saussurean characterization of the word *flow* would assert that there is no apparent reason for the choice of this particular signifier to represent [fluidity]; any other would have done the job just as effectively. Thus, *glop, plip, druck,* for instance, could have arbitrarily been chosen instead.

However, what such a characterization ignores is that even those who do not speak English, will notice an *attempt* in the above signifier to imitate the sound of moving water–an attempt constrained by the sound system of English. Users of this word no longer consciously experience it as a vocal simulacrum of [fluidity], because time and constant usage have made them forget its probable iconic origin. In theory, the word *flow* is, as Saussure correctly asserted, one of an infinite number of permissible combinations of English phonemes that can be envisioned to represent the [fluidity] property in question. But, in practice, it is implausible that *flow* was coined arbitrarily to accomplish this task. More than likely, the congener of that signifier *attempted* to simulate the sound s/he perceived moving water to make, with the phonemic resources of English. Now, whether or not this is what actually happened is besides the point–in actual fact, the word derives from Old High German *flouwen* 'to wash', which comes from the Indo-European base **pleu-* 'flood'. The relevant thing to note here is that once people are presented with this hypothetical "simulative scenario", they start typically to experience the signifier consciously as onomatopoeic, rejecting alternative candidates that could in theory have been chosen to refer to the [fluidity] property (e.g., *glop, plip, druck jurp, flim,* etc.) as somehow "unnatural". This kind of anecdotal evidence is rather extensive, strongly suggesting that word-creation is hardly an arbitrary, discretionary process, but rather, one that is guided initially by the PMS.

The linguist Ronald Langacker (e.g., 1987) has argued that iconicity is not just a characteristic of word-creation. He has claimed that even the parts of speech have an iconic referential function built into them. To grasp what this implies, take, for instance, the part of speech commonly called a *noun*. What is significant about nouns, Langacker suggests, is the fact that, by their very nature, they must refer to an implicit iconic property of the referents that they encode, namely, whether they have a boundary or not. It is this [bounded] feature that is built into the representational function of nouns. Nouns encoding [+bounded] referents are called *count*; those encoding [-bounded] ones are called *noncount*. Thus, for example, the noun *leaf* is classified as a count noun, because it evokes the image of a referent with a bounded region, whereas the noun *water* is classified as a mass noun, because it elicits the image of a referent with no foreseeable boundaries. Now, all this has specific grammatical consequences, as evidenced by the fact that the noun *leaf* can be pluralized–because leaves have boundaries and can thus be counted–whereas *water* cannot, unless the referential domain is metaphorical (*the waters of Babylon*); *leaf* can be preceded by an indefinite article *(a leaf)*, *water* cannot; etc. The [bounded] feature is also implicit in the forms created in nonverbal modeling systems to represent nominal referents–in painting, for instance, *water* is depicted as having either no boundaries or else as being bounded by other forms (land masses, the horizon, etc.); *leaves,* on the other hand, are depicted as separate figures not dependent upon such boundaries (see Langacker 1987, 1990).

2.2 Primary singularized modeling

Although Peirce gave the notion of *iconic sign* its formal enunciation, it manifests a type of representational phenomenon that has been known about since ancient times (Lausberg 1960: 554; Wells 1967). Indeed, the debate between naturalists and conventionalists in ancient Greece revolved around whether or not a word represented things naturally (iconically) or not. Today, this debate is viewed as fruitless within biosemiotics because of the extensionality principle (§1.2.1), which claims that abstract (conventionalized) forms are derivatives of more concrete, sense-based forms.

The difference between human and nonhuman primary modeling lies not only in the infinite range of signifieds that human iconic forms are capable of encoding, but also in the unique capacity of human beings to extend the "sense-inferred" properties reproduced by iconic forms in various ways. Consider, for instance, the verb *drop*. This was, no doubt, coined originally to simulate the sound made by an object falling to the ground, a referent which can be abbreviated to [falling sound]–in actual fact, this word derives from Old German *dreopan* 'to drop' which, in turn, can be traced back to the Indo-European base etymon **dhreub-* 'to break'. Now, through extensional modeling, this vocal icon can be applied to abstract referents or referential domains (connotata) that are felt, by extension, to involve [falling] or some signified that can be seen to be related to it in some specific way:

- [to utter (a suggestion, hint, etc.) casually]: e.g., *Drop* me a hint, if you can.
- [to send (a letter)]: e.g., *Drop* me a line or two, when you can.
- [to stop or dismiss]: e.g., *Drop* that case!
- [to lower or lessen]: e.g., She *dropped* some of her conditions
- [to omit (a letter or sound) in a word]: e.g., You always *drop* your vowels.
- [to leave (a person or thing) at a specified place]: e.g., I'll *drop* you off at home.
- [a minimal amount]: e.g., That is barely a *drop* in the bucket.
- [at the slightest provocation]: e.g., At the *drop* of a hat, she always reacts.
- [to retreat]: e.g., He *dropped* back.
- [to lag behind]: e.g., She has *dropped* behind.
- [to pay a casual or unexpected visit]: e.g., John always *drops* in unexpectedly.
- [to decrease]: e.g., There's been a *drop* in the Dow-Jones.
- [to leave]: e.g., They *dropped* out of the course, because they couldn't keep up.
- [to cease]: e.g. Just *drop* it; I'm not interested in what you are saying.

The modeling history of *drop* is shown in figure 14 below:

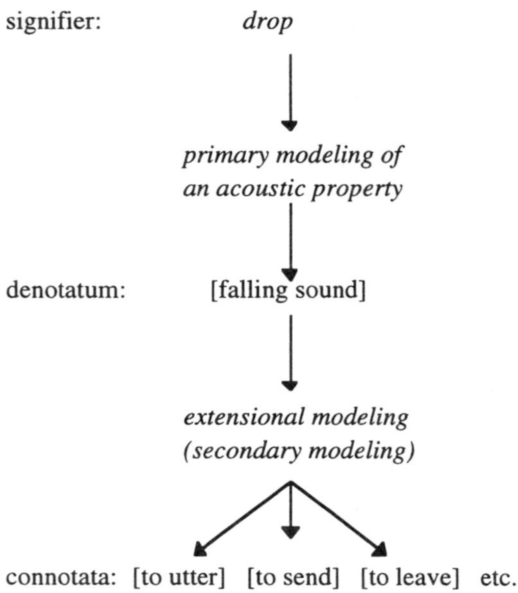

Figure 14. Modeling history of *drop*

This figure shows that after a sensory property of some referent or referential domain is encoded in a signifier, the resulting denotatum can be applied to referents that are felt to connote this same property, by extension.

2.2.1 Primary nonverbal singularized modeling

Primary nonverbal singularized modeling behavior is found throughout the zoosemiotic realm (Wickler 1968). A few illustrations will suffice here. The iconic function of insect chemical signaling, for instance, has been amply documented by ethologists:

- Fluctuations in the intensity of odor trails (= singularized forms) laid by successful foraging ant species are proportionate to the amount and quality of the source of nourishment (Butler 1970: 45).
- The walkingstick insect (*Phasmida*), which closely resembles the twigs of the plants on which it lives, has the

ability to change minute details (= singularized forms) of its physical appearance as required.

- The black chunky myna bird (*Sturnidae*) from tropical Asia and the East Indies has a white spot on its wing (= singularized form) and bright yellow wattles (= singularized forms) on the back of its head, which allow it to blend in with the visible features of its habitat.

Sometimes a creature even has the capacity to alter its very surroundings to fit its own image by fabricating a number of dummy copies of itself to misdirect predators away from its body, the live model, to one of the copies. This capacity is shown, for instance, by different species of a highly interesting genus of spiders known as orb-weavers (Wickler 1968; Hinton 1973).

A truly remarkable example of antipredation iconicity is found in the soft-bodied aphid species, which is extremely vulnerable to predator attack (Kloft 1959). The members of this species are protected by ants with which they communicate by an alarm pheromone that functions to stabilize their association. Their relationship is further reinforced by the fact that the ants "milk" the aphids by vibrating their antennae against their backs; the aphids then secrete droplets of honeydew which are consumed by the ants. Kloft (1959) has suggested that this congenial relationship rests on a "misunderstanding", and has proposed, as a working hypothesis, that the hind end of an aphid's abdomen, along with the kicking of its hind legs, constitute, for an ant worker, an iconic signifier, standing for the head of another ant together with its antennae movement. In other words, the ant probably identifies the replica (the rear end of the aphid) with the model (the front end of the ant), and then goes about its business on the basis of this misinformation.

Such multifarious manifestations of iconicity in the zoosemiotic realm led René Thom (1975: 72) to suggest that the principal role of the central nervous system of animals is to map out localized regions so that the organism can simulate its own position in its environment, as well as detect prey and predators. An animal is constantly informed and impelled by natural iconic signifiers designed to release pertinent motor reflexes, such as approach (say, toward a prey) or withdrawal (say, from a predator). Among animal behaviorists,

Schneirla (1965: 2) has argued persuasively that "operations which appropriately increase or decrease distance between organisms and stimulus sources must have been crucial for the survival of all animal types".

Manifestations of natural singularized iconicity are found in the human sphere as well: e.g., a person's shadow cast upon the ground, a shape reflected in water, etc. But manifestations of this sort are ordinarily devoid of semiotic value. Even though they are potential iconic signifiers, they attain representational value only under special circumstances. A shadow can be cast only if the model is illuminated by a luminous source, the light hitting it, thus defining its shadow. A specular image is similarly formed in the reflecting surface. In both cases the resulting image vanishes with the disappearance of either the model or the luminous source.

Most of human nonverbal iconic modeling is intentional (mimetic). Artificial scents, for instance, are manufactured by humans to simulate natural odors. Chemical food additives are produced in imitation of the flavors of natural foods. And the figures that people draw are typically intended to replicate pictorially real-world objects, events, etc. The latter talent emerges at about the same time that children utter their first words. Although children, with parental prompting, may learn to label the rough scribbles they make as "suns" or "faces", they do not set out to sketch anything in particular, but instead seem spontaneously to draw visual forms that become refined through practice into precise, repeatable shapes. The act of drawing in childhood appears to be pleasurable in itself; usually identification of the forms is provided, if at all, only after the child finishes drawing. Of course, shapes eventually suggest "things" (signifieds) to the child as h/er ability to use language develops, but in the beginning, the child appears to engage in drawing simply because it gives h/er pleasure and satisfaction.

2.2.2 Primary verbal singularized modeling

As mentioned above (§2.1.2), Saussure saw word-creation as a largely arbitrary process. For Saussure, onomatopoeia (*plop, smash, bang, zap*, etc.) was the exception in verbal semiosis, not the rule. Moreover, the highly variable nature of onomatopoeia across languages suggested to him that even this iconic phenomenon was, in practice, also subject to cultural conventions. For instance, the sound made by a rooster is rendered by the onomatopoeic signifier *cock-a-doodle-do* in English, but by a different one, *chicchirichí* (pronounced /kikkiriki/), in Italian; similarly, canine barking is rendered by *bow-wow* in English, but by *ouaoua* (pronounced /wawa/) in French. Saussure maintained that such cross-cultural comparisons showed rather conspicuously that onomatopoeic constructions were subject to conventionalized perceptions of sounds.

However, as we argued above, what Saussure seems to have ignored is that even those who do not speak English, French, or Italian will be able to see that the above signifiers were probably created as attempts to imitate the sounds that roosters and dogs are perceived to make–attempts constrained by the respective phonemic systems of the three languages that are, in part, responsible for the different simulative outcomes. Since many icons are understood mainly in cultural context, as Saussure correctly suggested, Peirce coined the term *hypoicon* to acknowledge this fact. A hypoicon is an iconic singularized form whose construction is constrained by the subsystems found in a specific language (phonemic, morphemic, etc.). Nevertheless, the primary modeling qualities of a hypoicon can be easily discerned by even those who do not know the language, once they are told what it means. As hypoicons, the words *cock-a-doodle-do* and *chicchirichì* can be easily perceived to be simulative of the sound a rooster makes, even by those who do not speak English or Italian, but are familiar with roosters; i.e., they can see an attempt in both these signifiers to model the sound made by a rooster, even if the end results have turned out to be different.

Vocal iconicity manifests itself in such common simulative phenomena as:

- alliteration, or the repetition of sounds, for various imitative effects: *ding-dong*; *no-no, zig-zag, bang-bang; pow-pow*; etc.;
- the lengthening of sounds for emphasis: "Yesssss!"; "Noooooo!"; etc.;

- the use of different voice tones to express emotional states: "Are you absolutely sure?" "No way!" "I really wouldn't say that, if I were you!" etc.;
- sound-mimetic constructions such as those that characterize the language of comic books: "Zap!", "Boom!", "Pow!", etc.;
- onomatopoeic descriptions of people and things: e.g., a snake is described as *slithery, slippery*, etc. in imitation of the sounds that snakes are perceived to make;
- an increase in the loudness of the vocal delivery of a message in order to convey a state of anger, excitement, etc.; and a decrease to convey the opposite (calmness, composure, etc.)
- an increased rate of speech to convey urgency, agitation, etc.; and a decreased rate to convey the opposite state (placidity, indolence, etc.).

The linguist Morris Swadesh (1971), who was a pioneer in the study of vocal iconicity, drew attention, moreover, to vocal osmotic phenomena that are rarely recognized as such. For example, he found that in many of the world's languages forms created with [i]-type vowels encoded referents marked by the feature [+nearness], while forms created with [a]- [o]- and [u]-type vowels encoded referents marked by the opposite feature of [-nearness]. This cross-lingual pattern suggested to him that the notion of [+nearness] tended to be simulated by the relative nearness of the lips in the articulation of [i] and other front vowels, while the opposite notion of [-nearness] tended to be simulated by the relative openness of the lips in the pronunciation of [a], [o], [u] and other mid and back vowels:

Table 2. Nearness and distance concepts in English

Nearness concepts	Distance concepts
here	there
near	far
this	that
etc.	etc.

2.2.3 Binary iconic features

Features such as [nearness] (§2.2.2) and [bounded] (§2.1.2) can be called *binary iconic features*. These stand for the sensory properties inferred in referents (= SIH). As we saw above (§2.1.2), these features are even built into the referential functions of the parts of speech (Langacker 1987, 1990; Heine 1998). When specific nouns, verbs, adjectives, etc. are created or used, they can be seen to encode referents in terms of some specific binary iconic feature. Thus, for example, count nouns such as *apple* encode referents marked by the feature [+bounded]; whereas noncount nouns such as *rice* encode referents marked by the feature [-bounded]. As we saw, this inbuilt iconic property of noun reference entails structural effects at different levels of grammar:

Table 3. The referentiality functions and derived grammatical features of count and noncount nouns

Count Nouns: e.g., *apple*	Noncount Nouns: e.g., *rice*	Referentiality Function and Derived Grammatical Features: [+bounded] vs. [-bounded]
can be pluralized: *apples*	cannot be pluralized	the referents of count nouns are characterized by a boundary in their shape that allows them to be counted (and hence their signifiers pluralized)
can be preceded by the indefinite article: *an apple*	cannot be preceded by the indefinite article	a count noun can be perceived as a distinct thing (and hence specified as such with *a* or *an*)
cannot be preceded by an indefinite pronoun	can be preceded by an indefinite pronoun: *some rice*	the referents of noncount nouns have no boundary and can only be specified in an indefinite way

As another example, consider the category of *adjective*. Some adjectives refer to qualities perceived in the referents that nouns encode: e.g., "The apple is *red*". This type of adjective is marked by the feature [+quality]. Other kinds of adjectives are not marked by this feature; they are marked, therefore, by its opposite [-quality]: e.g., the adjective *my* does not refer to any quality in the referents that nouns encode. The [quality] feature thus separates adjectives, referentially, into *descriptive* (such as *red*) and *nondescriptive* (such as *my*). This inbuilt iconic feature also entails specific structural effects at different levels of grammar. For instance, in English adjectives marked by [+quality] can be inflected *(red - redder, reddest,* etc.), those marked by [-quality] cannot. This is due to the fact that inflections allow for the encoding of "degrees" of a specific quality (in this case intensity of color) as detected in a referent: "My apple is *red*, but yours is *redder*".

2.3 Primary composite modeling

Primary composite modeling is a representational strategy whereby various iconic signifiers are combined in order to encode complex (non-unitary) referents. In the human semiosic realm, poetic texts, maps, and most diagrams are products of artificial primary composite modeling. Take, as an example, the following four-line "sound poem" whose first nine words begin with the cluster /sl/ in alliterative imitation of a snake:

Slimily, sluggishly, slithery,
Slowly, slyly, slippery,
Slothfully, sluggishly, slumberously,
A snake passes by.

This salient phonic feature of the text–/sl-/ = [serpentine sounds]–has allowed us to model the [snake] referent in a simulative composite manner. Now, by changing the final line of the poem, the same phonic feature can be extended to encompass human personality traits, such as [vileness], [untrustworthiness], etc.:

Slimily, sluggishly, slithery,
Slowly, slyly, slippery,
Slothfully, sluggishly, slumberously,
A supposed friend passes by.

The modeling history of this second text is shown in figure 15:

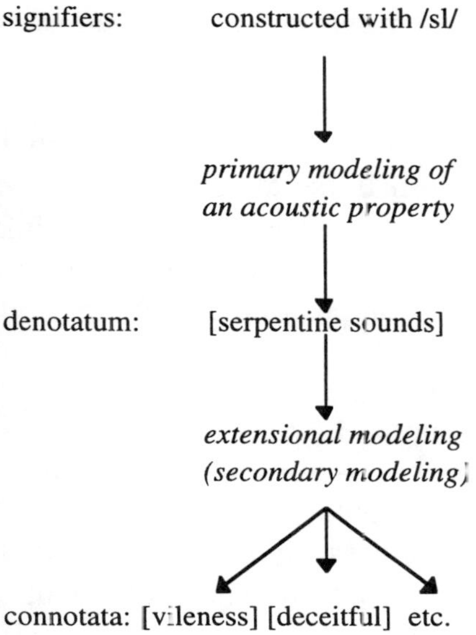

Figure 15. Modeling history of a poetic text

2.3.1 Primary nonverbal composite modeling

Manifestations of natural nonverbal primary composite modeling appear across species. As mentioned in the previous chapter (§1.4.2), a salient example of such modeling is the honeybee dance. After a bee discovers a new source of food, it fills its honey sac with nectar, returns to the hive, and performs a vigorous, highly patterned dance. If the source of food is within about 90 m. of the nest or hive, the bee performs a circular dance, first moving about 2 cm. or more, and then circling in the opposite direction.

The most conspicuous iconic feature about the performance of this dance is the fact that numerous bees in the hive closely follow the dancer, in imitation of its movements. Some then leave the hive and fly in widening circles, which are also iconic of the original dance, until they find the source. This type of simulative behavior is composite in the sense that it is mimics the set of movements of the dancer.

In regions where their territories overlap, some snakes mimic each other in a composite way in order to avoid the detection of their predators. The nonvenomous Sinaloan milk snake, for example, closely mimics the color pattern of alternating red and yellow rings of the venomous coral snake.

A widely-studied manifestation of natural primary nonverbal composite modeling in the human realm is facial expression. The psychologist Paul Ekman (1985), for example, has been able to link specific facial actions (signifiers) to different aspects of facial expressions. The signifiers that make up a composite facial expression include eyebrow position, eye shape, mouth shape, and nostril size. Ekman found very little variation across cultures in osmotic facial forms. Indeed, Ekman has shown that it is possible to write a "grammar" of the face that shows less cross-cultural variation than do language grammars. Feelings of [amusement], [anger], [surprise], and [sadness], to mention but a few, manifest themselves in facial forms that can be broken down into specific iconic constituents, which can, in effect, be defined as types of (non-pathological) symptoms of inner states.

The ability to make artificial primary nonverbal composite models is unique to anthroposemiosis. Take, as an example, the construction of a common *map*. A map is a composite indexical form in terms of its overall representational function, because it has been designed to indicate where real-world places are in terms of "map-world" places. However, the ways in which the places are laid out on the map reveal iconic modeling. Consider how a simple map is put together. Let's say that a stranger wants to get to a certain destination. The stranger is at location **A**, which is at the intersection of two streets, one running north and south, the other east and west. H/er objective is to get to a location **B**, which we know is two blocks west and three north of location **A**. An easy plan to show h/er how to get to **B** is to draw her a map. On the map, the location **A** can be shown as the point of intersection of two lines at right angles–standing iconically for the intersecting streets. Two equally-calibrated units added to the east-west line can then be added to represent two blocks west (to the left) of **A**, and three equally-calibrated units up from that point (i.e., in a northern direction) will show the desired location, **B**:

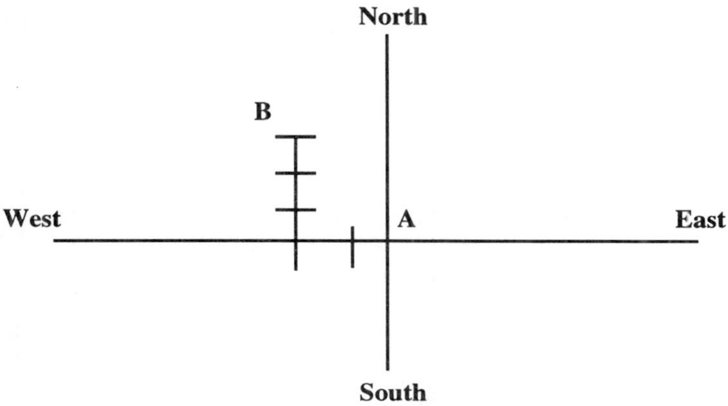

Figure 16. Map of a location **B** with respect to location **A**

By representing real-world spatial relations in such a way, the map allows its user to find an actual point in that real-world space through an iconic reconstruction process: i.e., through a mental conversion of the iconic visual relations displayed on the map-space into real-world geographical relations.

The use of maps, charts, and diagrams has greatly enhanced humankind's ability to know the world. These composite models are, in effect, elaborate *pictographs*–pictorial representations of scenes, events, ideas, etc.–allowing human beings literally to envision real-world phenomena in terms of iconic forms. The origin of pictographic modeling is ancient. The vivid carvings of animals that cover the roofs and walls of caves, such as those at Lascaux in France and Altamira in Spain, have been dated back some 30,000 years. These early drawings were the likely precursors of pictographs (Schmandt-Besserat 1978, 1989). As the hand movements used to make them became more abbreviated, they must have become more condensed and thus more abstract. Further abbreviations, for the sake of efficiency, must have led gradually to the practice of using a part of the pictograph to stand for a specific sound in the vocal word that also referred to the pictograph's signified. Sometimes called the *rebus principle*, this remarkable development led gradually to the invention of alphabets (see §3.1.3)–a *rebus* is a puzzle using words, pictures, or symbols to visually represent sounds that resemble the intended words of the solution. As psychologist Roger Brown (1986: 447-448) has aptly pointed out:

> It is usual to think of early writing itself as an independent invention of language, because the modern writing systems we know are almost all secondary to speech, and so necessarily came later than the spoken language. No alphabetic or syllabic writing system can be thought of as an instance of the invention of language. However, the earliest writing systems were all independent of speech and were not alphabetic or syllabic in character but were all pictorial or representational in character.

2.3.2 Primary verbal composite modeling

Needless to say, primary verbal composite modeling is unique to anthroposemiosis. The poem that we constructed above (§2.3) is an example of an iconic verbal text–i.e., of a primary verbal composite model composed, in part, with iconic signifiers. Originally fused with song, poetry was originally onomatopoeic in its overall representational character. In Western culture, poetry gained independence from singing in ancient times. In many other cultures, the two are still conceived as identical. The Italian philosopher Vico called the first speakers "poets", which etymologically means "makers", because, he claimed, their initial concepts were invariably encoded poetically (onomatopoeically and metaphorically). This original function has left its residues in the fact that, to this day, we continue composing many basic verbal texts poetically–from the rhymes we write for children, to the messages we put on greeting cards and to the jingles we create in advertising.

But primary verbal composite modeling is not only a feature of poetry. Its presence can be detected even in how the speakers of some languages organize the words they use in sentences. In some languages, called *agglutinative*, sentence-formation is largely independent of word order. In Classical Latin, for instance, the various relations among the words in a sentence were shown by inflection, i.e., by variations or changes that the words underwent to indicate their relations with other words. But, because of phonetic changes undergone by Latin words over the centuries, word order has taken over many of the relational functions of inflections in Latin's modern-day descendants (the Romance languages). In Latin, the sentence *Puer* ('the boy') *amat* ('loves') *puellam* ('the girl') could have been rendered in any one of six ways because the ending on each word would have told the interpreter of the sentence what relation each had to the others: *puer* is in the nominative case and is thus the subject of the sentence; *puellam* is in the accusative case (nominative = *puella*) and is thus the object of the sentence, no matter where it occurs in it:

Puer	*amat*	*puellam*
Puer	*puellam*	*amat*
Amat	*puer*	*puellam*
Amat	*puellam*	*puer*
Puellam	*amat*	*puer*
Puellam	*puer*	*amat*

In Italian, French, Spanish, or Portuguese, however, a different word order conveys a different meaning. In Italian, for instance, *Il ragazzo*('the boy') *ama* ('loves') *la ragazza* ('the girl') and *La ragazza* ('the girl') *ama* ('loves') *il ragazzo*('the boy') mean different things. The word order in Italian simulates the real-world sequence implied by the nature of actions: i.e., it mirrors the actual sequence that unfolds between an *actor*, an *action*, and the *acted upon*. In a phrase, the sequence of grammatical categories in the Italian sentence mirrors the structure of experience as it presents itself to consciousness:

[actor/agent] → [projects an action upon] → [something/someone]

↓	↓	↓
[subject]	[verb]	[object]
↓	↓	↓
Il ragazzo	*ama*	*la ragazza*

Figure 17. Iconic composite modeling of a real-world sequence

This sentence thus portrays, by its structure, the fact that the boy is the actor or perpetrator of an action which he projects upon the girl. Changing the order of the two encodes the reverse situation:

[actor/agent] → [projects an action upon] → [something/someone]

↓ ↓ ↓

[subject] [verb] [object]

↓ ↓ ↓

La ragazza *ama* *il ragazzo*

Figure 18. Iconic composite modeling of reverse real-world sequence

2.4 Primary cohesive modeling

A primary cohesive modeling system is a code consisting of particular types of iconic signifiers that serve various simulative representational purposes. An example of a primary code is the set of figures used commonly by engineers to make technical diagrams, such as those shown in figure 19 below:

Figure 19. Electrical circuits diagram

Although such diagrams are decoded symbolically, since their representational functions are products of conventional practices that must be learned in cultural context, the ways in which they are put together to represent the two types of circuits, [+open] and [-open], are clearly iconic.

Another example of a primary artificial code is the set of *stick figures* used commonly by artists to represent people. One well-known modern-day example of its use in visual artistic depiction is the cartoon strip *Cathy,* begun in the 1970s by the American cartoonist, Cathy Lee Guisewite (1950-), who used stick figures originally to record her feelings in a diary. From these she developed *Cathy* in 1976. The stick figures allow Guisewite to portray social relations in "bare outline"–which is, of course, the distinguishing iconic feature of stick figures themselves.

2.4.1 Primary nonverbal codes

Natural (osmotic) primary nonverbal codes are found across species. A remarkable example of such a code can be observed in the mound constructions that are engineered by common termites. These social insects have the ability to construct extremely hard walls from bits of soil cemented with saliva and baked by the sun. Inside the walls numerous chambers and galleries are constructed by these ingenious engineers, interconnected by a complex network of passageways. Ventilation and drainage are provided, and heat required for hatching the eggs is obtained from the fermentation of organic matter, which is stored in the chambers serving as nurseries. Of more than 55 species common in the United States, the majority build their nests underground. The subterranean termites are extremely destructive, because they tunnel their way to wooden structures, into which they burrow to obtain food. Now, upon close examination, the mound structure simulates the constituents of the termite's social evolution, even after the colony itself has become extinct–i.e., the mound visually mirrors the social organization of these architect insects. This is a dramatic example of unwitting cohesive iconicity manifesting itself in Nature as a property of a species' social behavior (Sebeok 1979).

Courtship and grooming codes across species are also fundamentally iconic. In most species of cockroaches, for instance, courtship unfolds with hissing noises, nibbling, and bobbing of the abdomen, in obvious simulation of sexual desire. In the human sphere, too, certain vocal tones and bodily signals during courtship displays are simulative of sexual interest and/or excitement.

Artificial nonverbal primary codes abound in human social life. Take, for example, the use of graphs. Graph-making systems constitute versatile codes for representing information and data in picture form, so as to be able to detect some pattern or trend in the information. The simple two-axial graph shown in figure 20 is a case-in-point. This shows the number of crystal glasses sold by a department store each day for a week. To find the number of glasses sold on day 3, all one has to do is locate the number 3 on the horizontal axis and then find the point directly above it. The position of this point vis-à-vis the vertical axis is 10, meaning that 10 glasses were sold on day 3. On day 4 the position of the point shows that 5 glasses were sold instead. The lower position of the dot on day 4 indicates a *drop* in sales. In effect, point locations *up* and *down* a graph mirror relative *increases* and *decreases* in the quantity of glasses sold. This is a salient iconic property of all graphs. Now, looking at the pattern of *ups* and *downs* of the points on the graph, we can detect an overall trend in sales: it would seem that sales of glasses started out at a fairly high level, dropped by mid week, and then picked up dramatically by week's end.

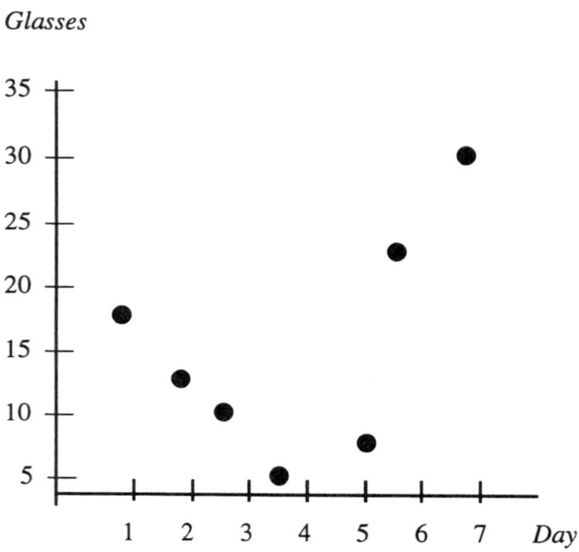

Figure 20. Two-axial graph

2.4.2 Gesture

A nonverbal code that has received comprehensive attention from semioticians and animal ethologists alike is *gesture*. Gesture can be defined simply as the use of the hands, the arms, and to a lesser extent, the head, to make bodily forms of all kinds (singularized, composite, etc.). Natural gestural codes are iconic by their very nature. For example, chimpanzees raise their arms in the air to indicate that they want to be groomed; they stretch out their arms to beg or invite; and so on.

Natural gestural iconicity in human semiosis is interconnected with vocal language. After videotaping a large sample of people as they spoke, the linguist David McNeill came, in fact, to the inescapable conclusion in 1992 that the gestures that accompany speech, which are called *gesticulants*, are hardly inconsequential to the act of vocal communication. Gesticulants exhibit images that cannot be shown overtly in speech, as well as images of what the speaker is thinking about. This suggests that vocal speech and gesture constitute a single integrated referential/communication system

McNeill proceeded to classify the gesticulants he observed into five main categories:

- *Iconic gesticulants*: As their name suggests, these bear a close resemblance to the referent or referential domain of an utterance: e.g., when describing a scene from a story in which a character bends a tree back to the ground, a speaker observed by McNeill appeared to grip something and pull it back. His gesture was, in effect, a visual icon of the action talked about, revealing both his memory of the scene and his point of view (he could have taken the part of the tree instead).
- *Metaphoric gesticulants*: These are, in effect, gestural externalizations of metaforms. For example, McNeill observed a male speaker announcing that what he had just seen was a cartoon, simultaneously raising up his hands as if offering his listener a kind of object. He was obviously not referring to the cartoon itself, but to the genre of the cartoon. His gesture represented this genre as if it were an object, placing it into an act of offering to the listener. This type of gesticulant typically accompanies externalizations of the so-called "conduit" metaform, [ideas = objects that are passed on along a conduit]: *presenting an idea, putting forth an idea, offering advice*, and so on.
- *Beat gesticulants*: These resemble the beating of musical tempo. The speaker's hand moves along with the rhythmic pulsation of speech, in the form of a simple flick of the hand or of finger movement up and down, or back and forth. Beats are indexes marking the introduction of

new characters, summarizing the action, introducing new themes, etc. during the utterance.

- *Cohesive gesticulants*. These serve to show how separate parts of an utterance are supposed to hold together. Beats emphasize sequentiality, cohesives globality. Cohesives can take iconic, metaphoric, or beat form. They unfold through a repetition of the same gesticulant form in the gesture space. It is the repetition itself that is meant to convey cohesiveness.
- *Deictic gesticulants*. These are aimed not at an existing physical place, but at an abstract concept that had occurred earlier in the conversation. They reveal a perception of concepts as having a physical location in space.

McNeill's gesticulants are called, more generally, *illustrators* because they can be seen to illustrate gesturally what is being said during vocal speech. In addition to illustrators, there are four other types of gesticulants: *emblems, affect displays, regulators,* and *adaptors:*

- *Emblems:* These directly translate words or phrases: e.g., the *Okay* sign, the *Come here* sign; the hitchhiking sign; waving; obscene gesturing; etc.
- *Affect Displays:* These communicate emotional meaning: e.g., the hand movements and facial expressions that accompany happiness, surprise, fear, anger, sadness, contempt, disgust, etc.
- *Regulators:* These regulate the speech of an interlocutor: e.g., hand movements indicating *Keep going, Slow down,* etc.
- *Adaptors:* These indicate or satisfy some emotional state or need: e.g., scratching one's head when puzzled; rubbing one's forehead when worried; and so on.

Human beings have also invented conventionalized gestural codes for various purposes. The gestural signifiers used by hearing-impaired people, by religious groups during periods of imposed silence or for various ritualistic practices, by music conductors, and so on, are all conventionalized forms which, nevertheless, manifest various degrees of iconicity. For example, in American Sign Language the statement "I stared at it for a long time" would be expressed as a single sign with the following iconic properties: (1) the signer forms [look at] by making a V under the eyes with the first and middle fingers of the right hand; (2) the hand moves out toward the object being looked at, repeatedly tracing an oval to indicate [over a long time].

Although there are cross-cultural similarities, substantial differences also exist both in the extent to which gesture is used and in the interpretations given to particular conventionalized gestural forms. In 1979, Desmond Morris, together with several of his associates at Oxford University, examined 20 social gestures in 40 different areas of Europe. The research team found, for instance, that many of the gestures had several meanings, depending on culture: e.g., the head gestures for [yes] and [no] used in the Balkans seem inverted to other Europeans; a tap on the side of the head can indicate completely opposite things–[stupidity] or [intelligence]–according to the society in which it was used; and so on.

2.4.3 Sentence structure

Needless to say, language is unique to human semiosis. In this section we will be concerned one last time with one specific iconic characteristic of this code–sentence structure–a feature of language that we have already discussed several times. In general, it can be said that word order in sentences tends to simulate real-world sequential or cause-and-effect phenomena: i.e., sentence structure largely mirrors the structure of experience as it presents itself to consciousness (Langacker 1987, 1990).

Consider, for instance, the relation between an active and passive sentence such as "Alexander is eating the grapes" vs. "The grapes are being eaten by Alexander". Upon close examination, it can be seen that the two sentences encode an iconic property of referents–their relative location to each other in terms of a mental [foreground]. In the active sentence, the *actor/agent* is in the [+foreground] (or in the [-background]) and the *acted-upon/receiver* in the [-foreground] (= [+background]). The *action* is spotlighted as an activity of the *agent* in the [+foreground] towards the *receiver* in the [-foreground]:

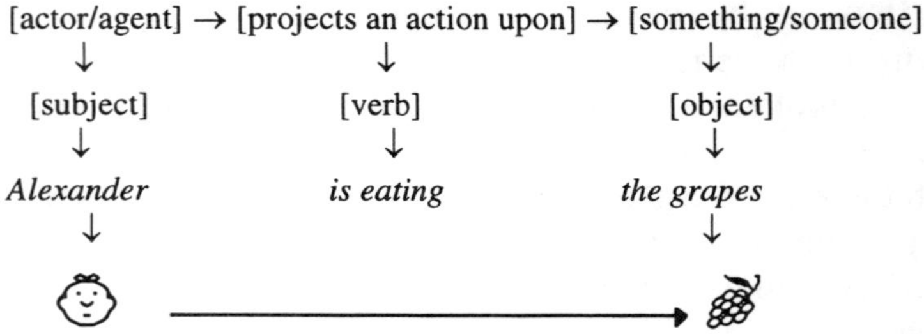

Figure 21. Iconic cohesive modeling of an active sentence

A change from passive to active, however, changes the arrangement of the [foreground] and the [background] to the mind's eye. By changing the grammatical position of the *receiver* and the *agent*, the passive sentence brings the *grapes* to the mental [+foreground], relegating the eater *Alexander* to the [-foreground]. The action of eating is now spotlighted on the receiver of the action, the *grapes*. The overall view that passive sentences encode, therefore, is one of the object as the receiver of the action, and the agent as being in the background (and thus less visible):

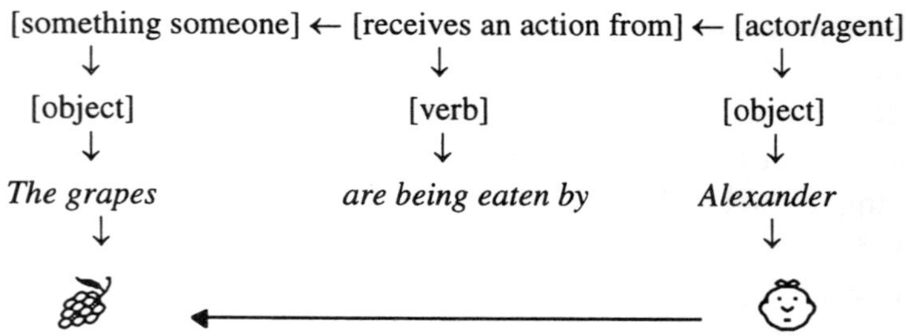

Figure 22. Iconic cohesive modeling of a passive sentence

The different mental views of the same event are built into the referential functions of these two sentence types. Sentences in which the agent occurs first grammatically are called, logically, *active*– because they evoke the image of an agent in the foreground acting on a receiver. Sentences in which the agent occurs last grammatically are called *passive*–because they elicit the image of a receiver of an action in the foreground with the agent in the background.

The [foreground] and [background] iconic features are also implicit in nonverbal modeling systems used to represent the two different types of sentences–in painting, for instance, a scene corresponding to the active sentence form would portray Alexander as the larger figure, catching the eye first as the perpetrator of the action; a scene corresponding to the passive sentence would highlight the grapes (through enlargement and positioning of the grape form), catching the eye first as the receiver of the action.

2.5 Primary connective modeling

A metaform is, by definition, a primary connective model. Take, as an example, the following metaphorical statements:

14. Those ideas are *circular*.
15. I don't see the *point* of your idea.
16. Her ideas are *central* to the discussion.
17. Their ideas are *diametrically* opposite.

These are clearly the result of a linkage between an abstract notion, [ideas], with a concrete source domain containing referents that can be easily seen or drawn, namely [geometrical figures/relations]. The iconicity implicit in this metaform can be seen in the nonverbal realm as well: e.g., in the common practice of representing ideas and theories with diagrams based on geometrical figures (points, lines, circles, boxes, etc.). All "theoretical models" are, in effect, geometric diagrams, externalizing the [ideas = geometrical figures/relations] metaform in some specific way.

Now, primary connective models, unlike all other kinds of models, never encode denotata. They encode only connotata. To see what this implies take, as an example, the metaform [human personality = perceived physical features of animals]. One of its externalizations can be seen in a sentence such as "The professor is a *snake*". Clearly, the meaning of [snake] that this statement encodes is not its denotative one–[serpentine reptile]–but rather, the complex of connotative meanings that we associate with snakes: namely, [slyness], [danger], [slipperiness], etc. Now, each different selection of a vehicle from the [perceived physical features of animals] source domain produces different connotata: e.g., in "The professor is a *gorilla*", the [professor] is conceived instead as someone [aggressive], [combative], [rude], etc.

The modeling history of this metaform is shown in figure 23 below:

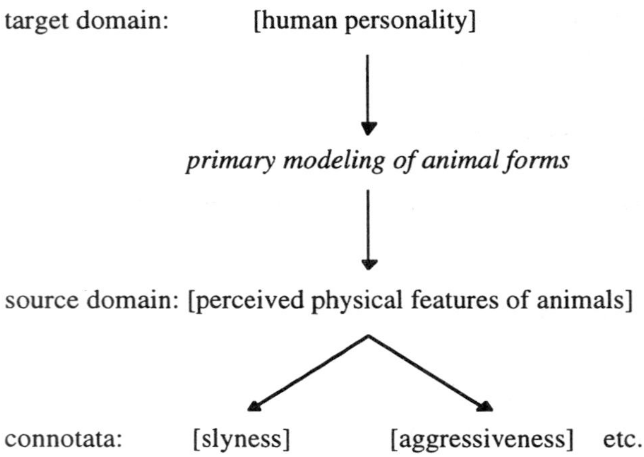

Figure 23. Modeling history of the metaform [human personality = perceived physical features of animals]

In primary connective modeling, therefore, it is not the denotative meaning of the vehicle (the item chosen from a source domain) that is transferred to the topic, but rather its connotata.

2.5.1 Metaforms

The probable reason underlying the crystallization of [human personality = perceived physical features of animals] metaform is the de facto perception that humans and animals are interconnected in the natural scheme of things. The probable motivation for the [thinking = seeing] metaform (§1.6.1) is the perception that thinking is "internal seeing". In effect, a metaform is the result of sense-inference

(§2.1.1)–the simulative process guided by the conscious inferences people make about the abstract referents they are attempting to encode.

The role of metaforms in abstract concept-formation has been amply documented by the relevant literature, which gained momentum in 1977 when Howard Pollio and his associates showed that metaphor was hardly a discourse option, but its very backbone (Pollio, Barlow, Fine, and Pollio 1977). This turning point led in the late 1970s and throughout the 1980s to the development of two significant trends: (1) conceptual metaphor theory itself (e.g., Ortony 1979; Honeck and Hoffman 1980; Lakoff and Johnson 1980, 1999; Lakoff 1987; Lakoff and Turner 1989; Kövecses 1986, 1988, 1990; Johnson 1987; Indurkhya 1992), and (2) a new branch of linguistics that now comes under the rubric of *cognitive linguistics* (Langacker 1987, 1990; Croft 1991; Deane 1992; Taylor 1995; Fauconnier 1997). The relevant research within conceptual metaphor theory strongly suggests that most of our abstract concepts are stored as metaforms by our memory systems.

As discussed in the previous chapter (§1.6.1), in a SA framework, a specific *metaphor* is not considered an isolated construction, but rather, a specific instantiation of a metaform. The following metaphorical statements, for instance, are all instantiations of the above [human personality = perceived physical features of animals] metaform:

18. The professor is a *snake*.
19. Keep away from my cousin; she's a *rat*.
20. What a *gorilla* my uncle has become!
21. She's a sweetheart, a true *pussycat*!
22. He keeps everything for himself; he's a real *hog*.

As mentioned, these instantiations produce various connotata–[slyness], [betrayal], [aggressiveness], [kindness], etc.–for evaluating specific personalities: i.e., each different selection of a vehicle from the source domain–[snake], [rat], [gorilla], [pussycat], [hog], etc.–provides a different connotative depiction of the specific personality to be evaluated. Needless to say, perceptions of animal behaviors vary according to situation and according to culture. But the fact remains that people the world over react experientially and affectively to animals in specific ways and that these form a source domain for evaluating human personality. The essence of connective modeling lies, in fact, in the *abduction* of properties from a source domain perceived as being interconnected with an abstract target domain:

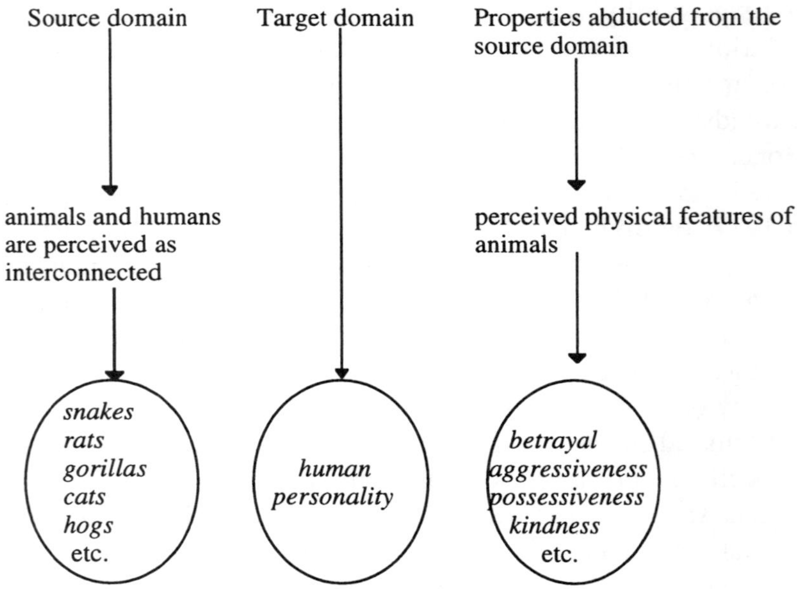

Figure 24. Connective modeling

Now, each specific selection of a vehicle from the domain becomes itself a source for providing further connotative detail for evaluating human personality, if such a need should arise. Thus, for instance, the specific choice of [snake] as the vehicle encompasses, further, its own sub-domain (made up of types of snakes), allowing one to zero in on specific details of the personality being described:

23. He's a *cobra*;
24. She's a *viper*.
25. Your friend is a *boa constrictor*.

In effect, source domains are further divisible into are sub-domains that provide the metaform-user with an array of connotata that s/he can utilize to project subtle descriptive detail on to a target domain. The metaform [human personality = perceived physical features of animals] itself corresponds to what psychologists call a *superordinate concept*, namely a concept with a highly general referential function (Rosch 1973a, 1973b). The choice of specific vehicles from the [animal] source domain–[snake], [rat], etc.–produces, in effect, what psychologists call *basic concepts*, namely concepts having a typological function. Finally, the selection of items from a sub-domain–[cobra], [viper], etc.–produces *subordinate concepts*, i.e., concepts that are needed for specialized purposes–this is why we describe someone as having *snake eyes*, if s/he is perceived as being a particularly dangerous individual; why we describe someone as being *chicken-livered*, if s/he is perceived as being an overly-cowardly individual; and so on.

Knowledge of human personality entails, in effect, knowledge of superordinate concepts such as the one discussed here. Clearly, this kind of knowledge is culture-specific. The very same source domain of [perceived physical features of animals] could have been utilized differently; i.e., applied to a different target domain such as [justice], [hope], etc. Or else, a different source domain could have been used in tandem with this metaform. In Western culture, for instance, the same target domain of [human personality] is frequently conceptualized in terms of [mask-wearing]. Indeed, the original meaning of the word *person* reveals this very conceptualization. In ancient Greece, the word *persona* signified a 'mask' worn by an actor on stage. Subsequently, it came to have the meaning of 'the personality of the mask-wearer'. This meaning still exists in the theater term *dramatis personae* 'cast of characters' (literally 'the persons of the drama'). Eventually, the word came to have its present meaning of 'living human being'. This diachronic analysis of *person* also explains why we continue to this day to use "theatrical" expressions such as *to play a role in life*, *to put on a proper face*, etc. in reference to persons, and why the metonymic metaform [the person = the face] is so pervasive in discourse:

26. What *face* should I put on today?
27. You must try your utmost to save *face* in that situation.

28. What *face* (effrontery) that person has!
29. You must confront her squarely, *face* to *face*
30. Why did you make that *face*?

It is interesting to note, briefly, that the above metaforms standing for personality underlie various symbolic practices and behavioral phenomena. The metonymic metaform [the person = the face], for instance, is the reason why portraits are based on the face, why people evaluate personality on the basis of facial appearance, etc. The face is, in effect, interpreted as a meta-symbol for personality (§1.6.2). And the metaform [human personality = perceived physical features of animals] is at the root of many common meta-symbolic practices: e.g., in totemism the practice of adopting an animal (or a plant, or a natural object) as the emblem of a clan or family; in the use of animal masks during certain festivities in cultures throughout the world; in the names given to sports teams (the Chicago *Bears*, the Detroit *Tigers*, etc.); in heraldic traditions; in the creation of fictional characters (*Bugs Bunny, Brer Rabbit*, etc.) for use in story-telling to children; in the creation of surnames (*Tom Wolf*; *Maxine Bear*; etc.); and so on.

2.5.2 Image schemas

Lakoff and Johnson (1980) trace the psychological source of metaforms to *image schemas*. These are, in effect, what we have termed above *binary iconic features* (§2.2.3), i.e., mental impressions of sensory experiences involving [location], [movement], etc. These mental templates assist us in explaining, mediating, or guiding our responses to perceptual inputs. They are, in other words, the internal iconic links between source domains and abstract concepts, permitting us to recognize patterns within certain perceptions, to anticipate certain consequences, and to make sense inferences and abductions. Schemas reduce a large quantity of sensory information into mental models, thus assigning a cognitive form to perception. In biosemiotic terms, they can be said to mediate between what von Uexküll (1909) called the *Umwelt* (outer world) and the *Innenwelt* (inner world). Image schemas are, in effect, mental outlines of things. They are not replicas; rather, they are images based on cultural norms and on personal experiences (Arnheim 1969).

The topic of imagery has a long history in psychology. Individual differences in the ability to experience imagery were recorded already in the previous century. The research that shows how mental imagery can be elicited is actually rather straightforward and, in our view, unambiguous. People can picture faces and voices accurately and quickly, rotate objects in their heads, locate imaginary places in their mind-space, scan game boards (like a checker board) in their minds, and so on, with no difficulty whatsoever. While researchers might disagree on exactly what it is that their subjects "see" or "experience" in their minds, there is general agreement that something is "going on" in the mind. Stephen Kosslyn (e.g., 1983), who is well known for having investigated empirically how the brain's imagery system might work, conducted series of ingenious experiments in the early 1980s that showed how subjects can easily conjure up images of the arrangement of furniture in a room, of how to move a couch, of how to redesign a blueprint, etc. Kosslyn's work demonstrated, in essence, how people construct elaborate mental images, search them out for specific purposes, and perform all kinds of imaginary movements.

Image schemas are so deeply rooted that we are hardly ever aware of their control over conceptualization. But they can always be elicited easily. If someone were to be asked to explain the expression "I'm feeling *up* today", s/he would not likely have a conscious image schema involving an upward orientation. However, if that same person were to be asked the following questions–"How far *up* do you feel?" "What do you mean by *up*?" etc.–then s/he would no doubt start to visualize the appropriate schema. In effect, image schemas are evidence of "abstractive seeing" as the philosopher Susanne Langer (1948) so aptly put it.

As an example of how image schemas guide the derivation of metaforms, consider the [impediment] schema (shown in figure 25 below). Several abstract scenarios are visualizable in terms of this schema: one can *go around* the impediment, *go over* it, *under* it, *through* it, or *remove* it and continue on towards the object. On the other hand, the impediment could successfully impede someone, so that s/he would have to *stop* at the impediment and *turn back*. All of these actions can be easily "seen" within mind-space.

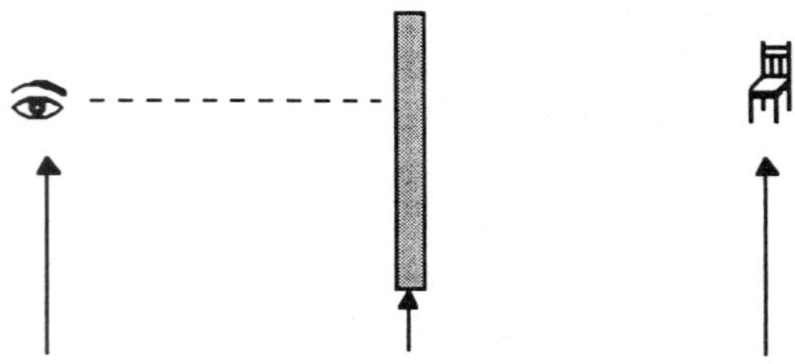

Figure 25. Impediment image schema

Now, the key thing to note is that this very image schema is the source for conceptualizing a host of abstract ideas:

31. With lots of determination, we *got through* that difficult time.
32. Jim felt better after he *got over* his cold.
33. There's no doubt that you will want to *steer clear of* financial debt.
34. With the bulk of the work *out of the way*, he was able to call it a day.
35. The rain *stopped* us from enjoying our picnic.
36. You cannot go any further with that relationship; you'll just have to *turn around* and *go back*.

Image schema theory suggests, clearly, that in primary connective modeling the source domains enlisted to deliver an abstract topic are not chosen in an arbitrary fashion, but rather, derived from the experience of events. Lakoff and Johnson (1980) have identified several basic types of image schemas. The three most discussed are called *orientational, ontological*, and *structural*. Orientational schemas are those that underlie the derivation of metaforms perceived to have "orientational structure"–[up], [down], [back], [front], [near], [far], etc. These undergird the formation of such abstract concepts as:

[mood]

37. I'm feeling *up* today.
38. She's feeling *down*.
39. That *boosted* my spirits.
40. My mood *sank* over our affair.
41. That gave me a *lift*.

[health]

42. I'm at the *peak* of my health.
43. She *fell* ill.
44. Getting over my disease has been a veritable *uphill* struggle.
45. She *dropped* dead suddenly.
46. His health is *sinking* fast.

[economy]

47. Inflation is *down*.
48. The economy is in an *upswing*.
49. The economy took a *downturn*.
50. The market has been in quite a *tumble* for a number of years.

[growth]

51. My income has gone *up*.
52. Your earning potential has *decreased* significantly over the years.
53. University standards are going *up*.
54. Job expectations have been *lowered* significantly over the past two years.

Ontological schemas underlie the derivation of metaforms perceived to have either the structural properties of [entities], [substances], [containers], [impediments], etc. or the characteristics of physical processes and forms–[plants], [movements], etc. Here are some examples of ontologically-based metaforms:

[the mind = a container]

55. I'm *full* of memories.
56. My mind is *empty*.
57. What's *inside* your mind?
58. Get that idea *out* of your mind.

[the mind = machine]

59. My mind is not *working*.
60. My memory is becoming *rusty*.
61. My mind is *coming apart*.
62. There are several *cogs* missing from my mind.

[knowledge = light]

63. I was *illuminated* by that professor.
64. I was *enlightened* by what happened.
65. That idea is very *clear*.
66. That theory is *brilliant*.
67. His example *shed light* on several matters.

[ideas = buildings]

68. That is a *well-constructed* theory.
69. His views are on *solid* ground.
70. That theory needs *support*.
71. Their viewpoint *collapsed* under criticism.
72. She put together the *framework* of a theory.

[ideas = plants]

73. Her ideas have come to *fruition*.
74. That's a *budding* theory.
75. His views have contemporary *offshoots*.
76. That is a *branch* of mathematics.

A structural image schema is one that combines ontological and/or orientational properties. This produces metaforms such as the following:

[time = a resource] + [time = a quantity]

77. My *time* is *money*
78. You cannot *buy* my *time*
79. His *time* is *valuable*.

[ideas = a commodity] + [ideas = quantities]

80. He certainly knows how to *package* his ideas.
81. That idea just won't *sell*.
82. There's no *market* for that idea.
83. That's a *worthless* idea.

People do not normally detect the presence of image schemas in such common expressions because of repeated usage. For example, we no longer interpret the word *see* in sentences such as "I don't *see* what you mean"; "Do you *see* what I'm saying?" in image schematic terms, because its use in such expressions has become so familiar to us. But the association between the biological act of seeing outside the body with the imaginary act of seeing within mind-space was originally the source of the metaform [thinking = seeing], discussed in the previous chapter (§1.6.1), which now permeates common discourse:

84. There is more to this than *meets the eye*.
85. I have a different *point of view*.
86. It all depends on how you *look at* it.
87. I take a *dim view* of the whole matter.
88. I never see *eye to eye* on things with you.
89. You have a different *worldview* than I do.
90. Your ideas have given me great *insight* into life.

Chapter III
Secondary modeling

Without words to objectify and categorize our sensations and place them in relation to one another, we cannot evolve a tradition of what is real in the world.

Ruth Hubbard (1924-)

3. Introductory remarks

As defined in the opening chapter (§1.2), the SMS is the system that undergirds both *indexical* (or *indicational)* and *extensional* modeling processes. The latter inheres in the ability to extend primary models both morphologically and connotatively for further representational uses. Extensional modeling is a uniquely human capacity, but the nonverbal form of indicational modeling has been documented in various species.

The objective of SA is, again, to document all manifestations of secondary modeling phenomena across species in order to derive general principles of semiosis in life forms, and then to examine the source and etiology of extensional modeling in the human species. The ability to extend primary forms to encompass abstract concepts is truly a remarkable achievement of human evolution.

3.1 The secondary modeling system

As mentioned, secondary forms are produced by *extensional* and *indicational* modeling strategies. The former entails the extension of primary singularized, composite, cohesive, or connective models into secondary ones, through connotation, morphological modification, or linkage (in the case of connective modeling).

A simple illustration of *extensional* modeling can be seen in how a word that was created to encode a sound property of some referent can, subsequently, be extended to encompass abstract referents. Take, for example, the English word *crash*. This was obviously coined initially as a primary singularized model simulating a [sudden shattering sound], as in: “The window *crashed* as the ball hit it”. Now, this very same signifier can be applied to abstract referents or referential domains that are felt, by connotative extension, to involve a [sudden shattering], as in: “Their business *crashed*”; “My computer system *crashed*”; etc. Moreover, it can be extended morphologically, with the addition of the suffix /-er/, to produce a new signifier, *crasher* “someone who crashes”. As this simple example shows in microcosm, language begets its representational power from the fact that it is largely an extensional modeling system, permitting human beings to encompass increasingly larger and more abstract domains of reference with a finite number of forms. In a manner of speaking, language is the system that extends the finite domain of sensory knowing into the infinite domain of reflective knowing.

Indicational modeling is secondary in the sense that it permits reference to things in terms of their spatiotemporal relation to other things. Thus, for example, while referring to a round object by joining the thumb and index fingers to represent a circular object constitutes a primary singularized model, referring to that object’s location with the index finger constitutes a secondary form of reference because it relates the object to its context of occurrence. Unlike icons, indexes are not substitutes for their referents.

3.1.1 Language vs. speech

It is crucial to distinguish between *language* and *speech* in any discussion of extensional modeling. As is widely known, Saussure (1916) used the analogy of a chess game to illustrate the difference between the two. To play a game of chess both players must first know the *language* of chess–i.e., the rules of movement of the chess pieces, the repertoire of response strategies to certain moves, etc. This system of abstract knowledge is gained through observation, instruction, and experience. Now, the actual choices made by a player during a game encounter constitute *chess speech*: i.e., the ability to apply the abstract knowledge of how chess is played to a specific game-playing situation. Knowing a language, likewise, constitutes abstract knowledge (which Saussure called *langue*);speech on the other hand entails the ability to apply this knowledge to specific situations (which Saussure called *parole*).

Requiring complex coordination between brain and vocal organs, it is baffling to consider why speech came about in the first place. The physiology of speech is dependent upon the larynx descending in the neck in early childhood (Laitman 1983, 1990). At birth, the position of the larynx in human infants is high in the neck, as it is in that of other primates. Infants breathe, swallow, and vocalize in ways that are physiologically similar to gorillas and chimps. But, during the first six months of life, the infant's larynx descends gradually down into the neck to make speech possible, at the same time dramatically altering the ways in which the child will carry out physiological functions from then on. Interestingly, research on the casts of human skulls has established that the lowering of the larynx originated around 100,000 years ago. This suggests that there may have been language without speech in pre-*Homo Sapiens* hominid species. The most probable mode of delivery of language was gestural. When speech became physiologically possible around 100,000 years ago, it is likely that it was used in tandem with gestural signs. However, it did not replace the latter completely. This is the most conceivable reason why we still use gesture as a default mode of communication (when vocal speech is impossible), and why we gesticulate when we speak (see chapter II, §2.4.2).

Another enigma of human evolution concerns the emergence of the complex neural structures that the brain had to develop in order to make possible the comprehension and production of speech. Speech is generated in Wernicke's area in the left hemisphere of the brain and then transmitted to Broca's area in the same hemisphere for encoding and, finally, on to the motor cortex which coordinates the appropriate articulatory organs for physical transmission. In speech comprehension, acoustic signals arrive in the auditory cortex from the ear and are then transferred to the adjacent Wernicke's area where they are interpreted semantically. According to the paleoanthropological record, early hominids had not evolved either the physical (laryngeal) or neurological (speech-processing areas of the left hemisphere) features required for speech delivery (e.g., Lieberman 1975, 1984, 1991). Thus, in early hominid groups *langue* probably existed without *parole*. Only after the evolution of the appropriate areas of the brain for speech, probably around 100,000 years ago, did it become possible for humans to deliver *langue* vocally. There is no evidence to suggest that earlier hominids either could speak or needed to communicate by purely vocal means. As mentioned, the endocranial research shows quite unambiguously that it was only around 100,000 years ago that *Homo Sapiens* had evolved the required anatomy for speech. The paleoanatomical study of hominid vocal tracts has also revealed that speech was developed at the expense of an anatomical system intended primarily for breathing and eating–modern humans can choke from food lodged in the larynx, other primates cannot. From an evolutionary perspective, this can only be explained by positing that the emergence of language was an *extension*–not an adaptive modification–of bodily functions. Endocranial casting has also made it possible to estimate the time frame of language evolution. *Homo Erectus* (1-1.5 million years ago) had a large brain (800-1300 cm^3). This hominid undoubtedly had *langue* (which it probably expressed through gesture), but not the neurophysiological capacity for *parole*. Actually, the ability to communicate in other ways, especially through gesture, may have originated quite back in time to *Homo Habilis* some two million years ago, a hominid species which had a relatively large brain (600-800 cm^3). The emergence of full speech, however, is traceable to *Homo Sapiens Sapiens* (40,000-100,000 years ago).

3.1.2 Extensional vs. indicational modeling

As discussed above, there are two types of extensional modeling strategies: (1) the process of extending signifiers through some modification of their physical forms, which can be called *morphological extensionality*; (2) the process of extending the meanings of primary forms to encompass connotative meanings, which can be called *connotative extensionality*. As an example of morphological extensionality, consider the word *logic*. This signifier can be modified by attaching two parts to it, called *affixes*, producing the new form *illogical*: the affix /il-/, known more specifically as a *prefix*, has a recurring functional meaning–[opposite of]); and the affix /-al/, known as a *suffix*, also has a recurring functional meaning–[the act or process of being something]. The form *illogical* incorporates the aggregate of meanings of the original form and the affixes. As an example of connotative extensionality, consider the word *square*, whose primary signified, [**B**], can be broken down into the following three distinctive features: [box figure], [four equal straight lines], [meeting at right angles]. These features allow us to determine if a specific real or imaginary figure under consideration will fall within the denotative scope of [**B**]. It is irrelevant if the lines are thick, dotted, 2 meters long, 80 meters long, or whatever. So long as the figure can be seen to have the distinctive features [box figure], [four equal straight lines], [meeting at right angles], it is identifiable denotatively as a *square*. Now, this same process of identification can be extended to encompass abstract referents, such as the following:

Table 4. Connotative extensions of the word *square*

Extended uses of *square*	Connotata
He gave me a *square* answer.	[honest, direct]
He is a *square* person.	[dull, rigidly conventional, out of touch with current trends]
That was a *square* deal.	[just, equitable]
They have *squared* their differences.	[settled]

An *honest, direct answer* is identifiable as a *square answer* because its [equalness] is as predictable as the actual square figure; a *conventional person* is identifiable as a *square person* because h/er personality is just as conventional or ordinary as a [box figure]; a *just deal* is identifiable as a *square deal* because it is an event that is [equal on all sides]; and *settling differences* is recognizable as *squaring differences*, since the process of making something *square* is equivalent to making things [equal on all sides].

Connotation can now be defined as the extension of a form [$\mathbf{A} \supset \mathbf{B}$] over a new meaning domain, [**C**], if [**C**] is recognized as entailing the features of [**B**] by implication. This can be shown as follows:

$$[\mathbf{A} \supset \mathbf{B}] \supset [\mathbf{C}] \Leftrightarrow [\mathbf{C} \supseteq \mathbf{B}]$$

This formula states that any primary form [$\mathbf{A} \supset \mathbf{B}$] can be applied to any other referent [**C**], if [**C**] entails [**B**] by implication ([**C**

$\supseteq$ **B**]). To use the above example of *square* connoting [justness] as a concrete case-in-point, the formula would be filled in as follows:

$$[\textit{square} \supset \text{four equal straight lines meeting at right angles}] \supset [\text{justness}] \Leftrightarrow [\text{justness} \supseteq \text{equality on all sides}]$$

Indicational modeling serves a different kind of referential function: it serves to direct attention to the location of referents in some contextual way. Indicational models specify such things as the [location], [situation], [presence], [absence], [distance], [occurrence], [directionality], [orientation], etc. of some referent with respect to other referents or to its context of occurrence. Take, for instance, the word *here*, as used, say, in the sentence "The ball is *here* in my hand". The indicational feature encoded by this word with reference to the *ball* is its [location]. This modeling history is shown in figure 26.

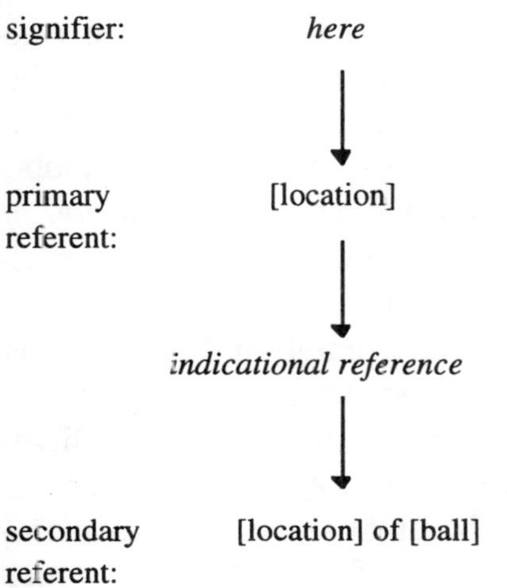

Figure 26. Modeling history of *here*

3.1.3 The alphabet: A case-in-point

Extensional modeling is at the root of many human abstract representational practices. A remarkable case-in-point of the handiwork of extensionality can be discerned in the origin of alphabets. Alphabets are, in effect, extensions of pictographic systems of representation. The starting point for explaining their evolution is around 100,000 years ago, when humans apparently developed the physiological capacity to speak (above §3.1.1). The temporal (hence linear) nature of vocal speech must have quickly become an important incremental communicative function, supplementing, in an intricate fashion, the entire human repertoire of pre-existing primary nonverbal modeling devices (such as gestural forms) already in place in the species.

At that point in time, pictographic representation must have merged, by simple association, with vocalization. Recall from the previous chapter (§2.3.1), the notion that pictographs–pictorial representations of scenes, events, ideas, etc.–most likely came about to allow for a "compressed" or "abbreviated" mode of drawing common objects. As the hand movements used to make drawings of, say, animals on cave walls became more abbreviated, they must have become more condensed and thus more abstract. Such abbreviated forms were, in effect, the first pictographic signifiers. Now, with the advent of vocal speech, further abbreviations, for the sake of efficiency, must have led gradually to the practice of using a part of the pictograph to stand for a specific sound in the vocal word used to encode the pictograph's referent. This remarkable event was, no doubt, the starting point for the development of *alphabetic representation*.

The alphabet is a secondary cohesive modeling system that allows humans to represent vocal singularized forms in terms of their constituent sounds. The earliest visual-representational systems, as we saw in the previous chapter, were all independent of speech. As early as the ancient civilization of Sumer around 3500 BC, pictographic representation was still used to record agricultural transactions and astronomical observations. Most of the Sumerian pictographs represented nouns; a few represented adjectival concepts such as [small], [big], and [bright]. A few centuries later, this pictographic system was expanded to include verbs: *to sleep*, for example, was represented by a person in a supine position. To facilitate the speed of writing, the Sumerians eventually streamlined their pictographs and transformed some of them into signs representing various sounds of speech. These were written down on clay tablets with a stylus in a form of writing known as *cuneiform*.

By about 3000 BC the Ancient Egyptians also used a primarily pictographic script, known as *hieroglyphic*, for recording hymns and prayers, registering the names and titles of individuals and deities, and annotating various community activities (*hieroglyphic* derives from Greek *hieros* 'holy' and *glyphein* 'to carve'). But, like the Sumerian pictographs, many of the Egyptian hieroglyphs eventually developed a partially symbolic function, standing for parts of words.

In a phrase, once pictographic representation became a flourishing enterprise in these ancient civilizations, it began to appear increasingly without the pictorial content, producing the first partial alphabetically-recorded forms. How could this have come about? The evolutionary transition from pictorial to alphabetic writing can be summarized somewhat (anecdotally) as follows. As the hand movements employed to make pictographs became abbreviated through constant usage, a more abstract form of visual representation emerged. So, for example, instead of drawing the full head of an ox, the following must have transpired:

Stage 1: Only its bare outline would have been drawn first (for the sake of efficiency).

Stage 2: This outline figure then came to stand for the [ox] through usage and/or by convention.

Stage 3: And this eventually came to stand for the vocal word for [ox] (*aleph* in Hebrew)

Stage 4: Finally, the figure, known as a *character*, came to stand just for the first sound in the word (*a* in *aleph*).

Stage (4) occurred around 1000 BC, when the ancient Phoenicians created the first true alphabet *characters* for referring to the consonant sounds of words. The Greeks adopted the Phoenician alphabet and called each character in it by such words as *alpha, beta, gamma*, etc., which were imitations of Phoenician words: *aleph* 'ox', *beth* 'house', *gimel* 'camel', etc. The Greeks then introduced symbols for vowel sounds, thus producing the first true alphabet, in the modern sense of the word.

Alphabetic representation is a truly remarkable achievement. It has made possible the recording and transmission of knowledge across time and across space, because alphabetically-recorded texts can be preserved in relatively permanent ways. In Western culture, to be an alphabet-user is to be *literate* and thus *educated*. The first schools in the West were a logical outgrowth of the invention of cuneiform writing. So close has the link between the two been forged that today we can scarcely think of knowledge unless it is recorded in some alphabetic form and preserved in some book or some computer data bank for posterity. But in every alphabetic symbol there is an iconic history and prehistory that has become "dim" because our eyes are no longer accustomed to extracting pictorial content from it.

3.2 Secondary singularized modeling

Secondary singularized forms are products of either extensional (morphological or connotative) or indicational modeling processes. The morphological extension of singularized forms, such as *illogical* (above §3.1.2), does not generate new signifieds. Rather, it allows form-users to modify the meaning of a form in regular ways. Take, for example, the morphological extension of the word *view* through prefixation, as in *review* and *preview*. The addition of *re-* adds the notion of [again] to the signified [seeing] encoded by the signifier *view*; the addition of *pre-* adds the notion of [in advance] to the signified:

view	=	[seeing]
review	=	[seeing again]
preview	=	[seeing in advance]

As we have discussed several times in this and in previous chapters, connotative extensional modeling testifies to the propensity of human beings to extend primary forms to encompass abstract referents. For example, a word such as *flow* (chapter II, §2.1.2) was no doubt coined iconically to simulate the sound made by the movement of water ("That river always *flows* smoothly") and of fluid pouring forth ("Sap *flowed* from the gash in the tree"). The distinctive features–[continuous], [smooth], [movement]–encoded by this word provide a "signifying template" for representing certain abstract referents:

Table 5. Connotative extensions of the word *flow*

Extended uses of *flow*	Connotata
Traffic *flowed* through the tunnel.	[moving with a continual shifting of the components]
The preparations *flowed* smoothly.	[proceeding steadily and easily]
The cadence of the poem *flowed* gracefully.	[exhibiting a smooth or graceful continuity]
Several conclusions *flow* from this hypothesis.	[deriving]
The book produced a *flow* of ideas.	[continuous outpouring]

The modeling history of this word is shown in figure 27.

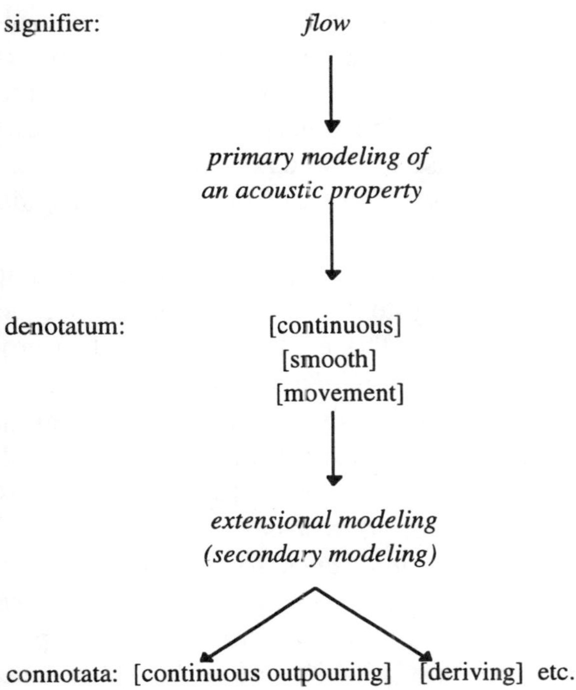

Figure 27. Modeling history of *flow*

Indicational singularized models are constructed to refer to things in terms of their location in time, space, or in relation to other things: e.g., a pointing index finger, a demonstrative word like *this* or *that*, or an adverb like *here* or *there* are locative indexes, serving to represent the relative spatial location of referents; adverbs like *before*, *after*, *now*, or *then* are temporal indexes, serving to represent the relative temporal relations that exist among referents; and pronouns like *I, you, he, she*, etc. are personal indexes, serving to identify the participants in a situation.

3.2.1 The extension of word forms and meanings

As singularized forms, words are constructed with the phonemes of a language. Recall from the opening chapter (§1.2.2) that a *phoneme* is the minimal unit of sound that allows speakers of a language to recognize words. This recognition effect is accomplished by means of binary oppositions between sounds. The words *sip* and *zip*, for instance, are kept distinct by a *binary* pattern of difference in their initial sounds. Both sounds are articulated in exactly the same way except in one detail: the former is a voiceless consonant ([-voice]); the latter a voiced consonant ([+voice]). This binary opposition–[-voice] vs. [+voice]–is *distinctive* in the case of /s/ and /z/, allowing speakers of English to perceive a *meaningful difference* in the two forms, *sip* and *zip*.

The morphological extension of words involves alterations that are consistent with the phonemic system. Take, as an example, the words *correct, logical,* and *rational*. These can be modified to produce the extended forms *incorrect, illogical* and *irrational*. All three prefixes have the same meaning–[opposite of]–but their different phonetic forms–/in-/, /il-/, and /ir-/–are not random or disconnected. In actual fact, it is the prefix /in-/ that stands for [opposite of]. This form can be added without change to words beginning with /n/ (*innumerable*), /t/ (*intake*), or /d/ (*indecent*), for example; but to words beginning with /m/ (*immature*), /l/ (*illegal*), or /r/ (*irregular*), on the other hand, the /n/ of the prefix /in-/ is assimilated to the phoneme that follows it: i.e., it becomes bilabial before /m/, lateral before /l/, and so on.

Morphological extension produces two kinds of modifications to the meaning and function of singularized forms. Consider the lexical item *learn*, which denotes [to gain knowledge or skill]. This signifier can be altered morphologically to produce extended forms such as *learned* and *learning*. The former has been modified by means of the addition of the suffix /-ed/. Like the prefixes /re-/ and /pre-/ discussed above (§3.2), which add the meanings [again] and [in advance] to the core meanings of words, the /-ed/ adds the signified [past] to the meaning of *learn*–namely that the action of learning has occurred in the past. The addition of the grammatical suffix /-ing/, on the other hand, produces a different kind of change: it creates a word with a different grammatical status than the word to which it is bound, as in "She is a woman of great *learning*": the word *learn* is a verb, while *learning* in the latter example is a noun. Morphological extensionality can be found in the nonverbal domain as well. For example, the figure of a cross, which can be used to stand for a crossroads on a traffic sign (§1.3.3), can be modified slightly by adding an arrowhead to it to expand its representational force (since the arrowhead emphasizes the fact that the crossroads is just ahead). The referential scope of the same cross figure can also be changed to indicate an emergency health service (e.g., the *red cross*) if it is painted red.

Extensional modeling comes naturally to humans. When children "play with words", as the expression goes, they are in effect using their SMS to extend words in creative ways. When the child reaches six to nine months, s/he starts to emit monosyllabic utterances (*mu, ma, da, di*, etc.), which are imitations of words the child has heard in social context. These are called *holophrastic* (one-word) utterances. They have been shown to serve three basic functions: (1) naming an object or event; (2) expressing an action or a desire for some action; (3) conveying an emotional state. Holophrases are typically monosyllabic reductions of adult words–*da* for *dog*, *ca* for *cat*, etc. Research has shown that over 60% will develop into nouns; and 20% will become verbs. During the second year children typically double their holophrases–*wowo* "water", *bubu* "bottle", *mama* "mother", etc. These early efforts are, clearly, extended forms that children make-up when they play to accompany their rhythmic movements, to simulate the sounds of their toys, and to refer to things.

The second main type of extensional process is connotative (§3.1.2). This is the process whereby the primary signified of a singularized form is extended to encompass abstract referents. For instance, the word *crash* (§3.1) can be extended to encompass such referents as [a sudden economic collapse] ("The market *crashed* yesterday") and [a system breakdown] ("My computer *crashed* yesterday"). Connotative extensionality can be found in nonverbal areas of representation as well. For example, the "V" figure made with the index and middle fingers to represent the letter *v*, can be extended to encompass abstractions such as [victory] and [peace].

Consider one other example of connotative extension. Recall from chapter I (§1.1.3) that the word *house* denotes [any (free-standing) structure intended for human habitation]. This meaning can be seen in utterances such as "I bought a new *house* yesterday", "*House* prices are continually going up in this city", "We repainted our *house* the other day", and so on. The signified, [**B**], of this word can be broken down into the following three distinctive features: [structure], [human], [habitation]. These features allow us to determine if a specific real or imaginary referent under consideration will fall within the denotative scope of [**B**]. Now, this same process of identification can be extended to encompass abstract referents as follows:

Table 6. Connotative extensions of the word *house*

Extended uses of *house*	Connotata
The *house* is in session.	[legislative assembly, quorum]
The *house* roared with laughter.	[audience in a theater]
They sleep at one of the *houses* at Harvard University.	[dormitory]

As mentioned above (§3.1.2), the distinctive features of [**B**]–[structure], [human], [habitation]–are implicit in such extensional uses. Any connotative extension of the word *house* is thus constrained by the distinctive features of [**B**]; i.e., *house* can be applied to refer to anything, [**C**], that implicates humans coming together for some specific reason.

[*house* (**A**) ⊃ structure for human habitation (**B**)] ⊃
[legislative assembly], [audience], [dormitory] (**C**) ⇔
[legislative assembly], [audience], [dormitory] (**C**) ⊇ structures involving humans in an occupied (inhabited) space (**B**)]

3.2.2 Indicational singularized modeling

Indexes are secondary models in the sense that they refer, as Peirce (II: 558) observed, to an "object not so much because of any similarity or analogy with it, nor because it is associated with general characters which that object happens to possess, because it is in dynamic (including spatial) connection both with the individual object, on the one hand, and with the senses or memory of the person whom it serves as a sign, on the other hand". Indicational modeling hinges upon association by contiguity.

While extensional modeling is unique to anthroposemiosis, natural indicational (indexical) modeling is found across species. Natural indexes occur in their most primitive form on the single-cell level, as physical or chemical entities, external or internal with respect to the embedding organism as a reference frame. Such indexes, which may be as simple as a change in magnitude, a mere shape, a geometric change in surface area, or some singularity, can be significant to a cell because they stimulate memory, exposing previously concealed information.

The ubiquitous prokaryotic bacterium *E. coli*–a bacillus normally found in the human gastrointestinal tract and existing as numerous strains, some of which are responsible for diarrheal diseases–provides a striking example of this form of natural indexicality (Berg 1976). This single-celled creature has multiple flagellae that it can rotate either clockwise or counterclockwise. When its flagellae rotate clockwise, they fly apart, causing the organism to tumble. When they are rotated counterclockwise, they are drawn together into a bundle that acts as a propeller to produce smooth, directed swimming. Roaming about in the intestinal tract, the bacillus explores a chemical field for nutrients by alternating–its context serving as operator–

between tumbling and directed swimming until it finds an optimally appropriate concentration of chemical attractant, such as sugar or an amino acid, for its replication. In doing so, it relies on a memory trace lasting approximately four seconds, allowing it to compare, over short periods of time and distances, where it *was* with where it *is*. On that basis, it "decides", with seeming intentionality, whether to tumble, stay in place, or swim and search for another indexical match somewhere else.

Another striking example of natural indexicality can be seen in the behavior of a small family of cerophagous picarian birds (*Indicator indicator*). This species has developed a remarkable symbiotic relationship with certain mammals–ratels, baboons, and humans–by employment of a purely indexical link, namely a strategy that the bird employs to guide its symbionts to the vicinity of wild bees' nests. A would-be guiding bird will come to, say, a person, and chatter until followed, but keep out of sight of the pursuer most of the time. Although its dipping flight is conspicuous, with the bird's white tail feathers spread out widely, the honey-guide "indicates" mainly by means of a repetitive series of chirps that subside only when it sees or hears flying bees whose nests, of course, are the target (Friedmann 1965).

The cellular processes and anatomies of organisms across Nature put on display an immense inventory of more or less obvious indexical markers of Selfhood. Krampen (1981) also found that, in phytosemiosis, indexicality predominates over iconicity. On the vegetative level, this would seem to correspond "to the sensing and regulating, in a feedback cycle, of meaningful stimulation directly contiguous to the form of the plant" (Krampen 1981: 195-196).

The human immune system utilizes approximately as large a number of cells dispersed throughout the body as the number of cells that composes a human brain. These endosymbiotic aggregations of spirochetal remnants (Jerne 1985) constitute an extremely sensitive, sophisticated repertoire of indexical signals, circumscribing, under normal conditions, Selfhood. Sadly, this indexical model of Selfhood can go awry under pathological conditions, when, for instance, one is afflicted with certain types of carcinoma, an autoimmune disease, or ultimately even when administered immuno-suppressors after an organ transplant.

In the human domain, artificial indicational singularized modeling is found throughout the world for the social purpose of *identifying* people. Handwriting authentication and fingerprinting (Moenssens 1971), for example, allow for the identification of people on the basis of specific traits. It is interesting to note that, in 1894, Mark Twain's fictional character *Pudd'nhead Wilson* became the first fictional lawyer in the world to use fingerprints in a criminal case, antedating Scotland Yard by eight years. Indeed, the whole field of forensic investigation constitutes, in its essence, a semiotic form of inquiry. Forensic science traces its roots to twelfth century England, when the office of coroner was created to keep a record of all criminal matters. American colonists brought the coroner system with them. In 1877, Massachusetts adopted a statewide system of criminal investigation headed by a coroner. Soon, other states adopted this arrangement. The coroner, or medical examiner, is, in effect, a semiotician who collects indexical signifiers, among other things, at the crime scene, conducts an autopsy in cases of death, examines the medical evidence and laboratory reports, and puts all this information together in a report. The forensic examination of substances found at a crime scene can often establish the presence of the suspect at the scene. Indexical identifiers present at a crime scene may include fingerprints, blood, hair, skin, or semen. Since 1985, forensic methodology has been extended to encompass DNA typing, a sophisticated composite form of indexical identification (see below §3.3.1).

3.3 Secondary composite modeling

Like singularized models, composite models can also be extended morphologically and connotatively. Consider maps, once again (chapter I, §1.4.3). A map is definable, overall, as an artificial *indexical composite form*. But it is not solely indexical; it represents places in topographical relation to each other according to some scale (iconicity), and it involves the use of legends (symbolicity). Now, a map can be extended morphologically to produce different "versions" of the same map-space: e.g., political boundaries, such as the limits of towns, countries, and states, can be added to the topographical map; landmarks such as tall buildings or prominent peaks on which a navigator may wish to take a bearing can also be added; etc. A cartographer may also devise a great variety of devices to suit various needs: e.g., a dot may be used to represent the presence of 10,000 head of cattle, or crossed pickaxes may be used to denote the location of a mine; etc. Such additions can both expand and change the referential scope of maps.

For the representation of the entire surface of the earth without any kind of distortion, a map must have a spherical surface; a map of this kind is known as a *globe*. A flat map cannot accurately represent the rounded surface of the earth except for very small areas where the curvature is negligible. To show large portions of the earth's surface or to show areas of medium size with accuracy, the map must be drawn in such a way that produces distortions of areas, distances, and direction. In some cases the cartographer may wish to achieve accuracy in one of these areas at the expense of distortion in the others. The various methods of preparing a flat map of the earth's surface are known as *projections*.

The process of projection produces connotata. The technique of *cylindrical projection* in Western map-making, for instance, produces various culture-specific connotata. Developed by the Flemish geographer Gerardus Mercator (1512-1594), this technique consists in wrapping a cylinder around the globe, making it touch the equator, and then projecting the lines of latitude outward from the globe onto the cylinder as lines parallel to the equator, and the lines of longitude outward onto the cylinder as lines parallel to the prime meridian (the line that is designated 0° longitude passing through the original site of the Royal Greenwich Observatory in England). The resulting two-dimensional map represents the world's surface as a rectangle with parallel lines of latitude and parallel lines of longitude (which are perpendicular to those of latitude):

Now, because of the curvature of the globe, the latitude lines on the map nearest the poles appear closer together. This distortion makes the sizes of certain land masses appear smaller than the sizes of other land masses. This entails various connotata associated with the represented areas: e.g., *larger land mass = better, more powerful, more important land mass*. Every type of traditional map produces its own kinds of connotata.

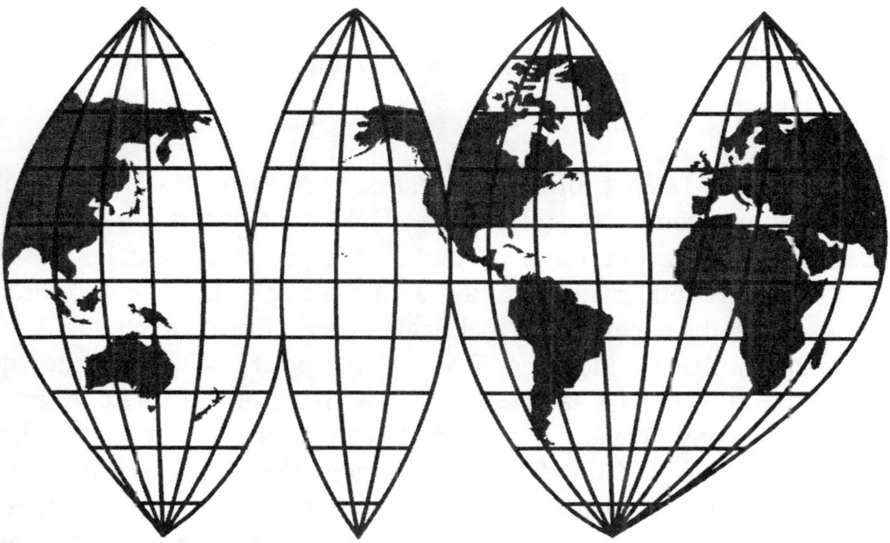

Figure 28. Projection map of the earth

To avoid this, recently a number of so-called computer projections have been developed mathematically for the accurate delineation of large areas on a small scale. Maps based on mathematical computation represent the entire earth in circles, ovals, or other shapes. For special purposes the earth often is drawn not within the original form of the projection but within irregular, joined parts.

3.3.1 Indicational and extensional composite modeling

The making of secondary composite forms is guided by the same principles that undergird the construction of singularized ones. In addition to maps, other common examples of artificial indexical composite forms are:

- the index sections put at the end of books which allow readers to locate certain topics and people in the book
- the flowchart diagrams employed in many fields to indicate the stepwise procedures used in performing a task, as in manufacturing, or solving a problem

- directories of all sorts, which identify addresses, phone numbers, etc.
- time-line diagrams, which portray temporal relations
- DNA profiles

So-called DNA profiling diagrams, used in forensic investigations, require some explication here. DNA profiling is a method of identification that compares fragments of deoxyribonucleic acid (DNA), the genetic material found within the cell nuclei of all living things. With the exception of identical siblings, the complete DNA of each individual is unique. A DNA "fingerprint" is constructed by first extracting a DNA sample from body tissue or fluid. The sample is then segmented using enzymes. The segments are marked with probes and exposed on X-ray film, where they form a pattern of black bars. These constitute the DNA fingerprint. If the DNA fingerprints produced from two different samples match, the samples are seen as coming from the same person.

Indicational representation is also a feature of language. There is a vast, separate literature devoted to the manifestations of indexicality in grammar, although it is not always recognized as such (Levelt 1989: 44-58). Karl Bühler (1934: 107), for instance, called the relevant context of the utterance *Zeigfeld*, or indexical field, and the anchoring point of this field its *Origo*, or origin (see also Jarvella and Klein 1982). Such composite indexicality can vary considerably from language to language and often be very complex in structure (Wills 1990). In one examination of the typological characteristics of personal pronouns in 71 languages, it was found that personal pronoun systems vary considerably in quantity, ranging from four to fifteen person indexes (Ingram 1978). In this array, the English five-person system (*I, you, he/she, we, they*) is highly atypical.

In a remarkable study of a single four-word sentence consisting of a modal auxiliary, a person-deictic pronoun, a verb, and the verb's complement, Fillmore (1973) hinted at the incredible intricacy demanded of a linguistic theory if it is to adequately capture the indexical properties of even the simplest sentences. To put it succinctly, a large measure of how language generates its forms and encompasses large domains of meaning depends decisively on indexicality.

Needless to say, any verbal text (a narrative, a play, a conversation, etc.) can be extended morphologically and connotatively. Take, for example, a weather forecast. This type of text is designed to provide information on probable meteorological conditions in the immediate future. It can be extended to encompass not only weather, but also the physical effects of the weather by adding UV (ultraviolet) and air pollution information to the report. Such extensions expand the information contents of the original text, not alter it referentially.

Fictional texts (novels, comic strips, films, etc.) are, by definition, connotative composite forms relating events in time and space that are modeled on real-life events. People interpret such texts, not as literal recountings of events, but as implying various psychological, sociological, or metaphysical meanings.

Literary historians and cultural semioticians have focused their energies on how a fictional text generates such connotata. First, it is obvious that the *plot*, *characters*, *setting(s)*, and *narrator(s)* of the text are invariably connotative in their referential scope: i.e., they refer to actions, people, places, etc. that are imaginary, and thus modeled by extension on existing actions, people, places, etc. The *plot* is basically what the text is all about; *character* refers to the portrayal of the people who are the primary participants in the plot; *setting* is the location where, and the time when, the plot takes place; and the *narrator* is the teller of the story. The narrator can be a personage of the narrative, the author of the narrative, or some other person or medium. Each type of narrator provides a different *perspective* of the story for the reader. The reader can thus feel a part of the narrative, looking at the action as if s/he were in it (*looking from within*); or aloof from it, looking at the action as if from the outside (*looking from without*). The novel *Flatland: A Romance of Many Dimensions*, written by the literary critic Edwin A. Abbott (1838-1926), is a noteworthy example of a narrative that provides the reader, literally, with both perspectives. The characters of the novel are personified geometrical figures, known as Flatlanders, living in a two-dimensional universe called Flatland. Flatlanders can only see each other as dots or lines, even if they are, from our vantage point, circles, squares, triangles, etc. The novel provides the reader with this perspective by juxtaposing h/er into the mind of a Flatlander. To grasp what kind of perspectival view this entails, Flatland can be imagined as the flat surface of a table. If one were an inhabitant in the surface, one would only be able to see figures in one or two dimensions: i.e., as dots or lines depending on their orientation (*looking from within*). For example, if one look at a circle paper cut-out lying on a table with one's eyes flush along the table's surface, one will see the circular cut-out as a line. The same applies to any other kind of cut-out—lines can also be seen as points if their orientation is such that they form right angles to the line of sight. The only way, then, to distinguish a circle from a straight line, an ellipse, or other figure is to view Flatland from a vantage point above it, i.e., to look down at the cut-outs from above the table. This three-dimensional viewing of the figures constitutes a *looking from without* perspective. It gives one literally a different *view* of Flatland and its inhabitants–a view with which the novel also provides the reader. Similarly, although the perspective in most other novels is not purely physical as it in *Flatland*, the reader's understanding of any narrative is invariably conditioned by one of these two mental vantage points–*looking from within* and *looking from without*. If the story is told by a character within it, then the view is from within. If it is told by an anonymous narrator, then the view is from without.

The serious study of fiction was initiated in semiotics after the Russian scholar Vladimir Propp (1928) argued persuasively that a large portion of ordinary discourse was built upon the same strategies used in modeling fictional plots, characters, and settings. According to Propp, there exists a relatively small number of "narrative units", or plot themes, which go into the make-up of an unconscious "plot grammar". It is this grammar that undergirds fictional texts and most conversations equally. After Propp, the semiotician who most influenced the study of narrativity in fiction and conversation, and their relation to language grammar, was the French scholar Algirdas Julien Greimas (1917-1992). Greimas's main contention was that human beings living in different cultures invent remarkably similar narratives with virtually the same stock of prototypical actions (plots), characters, and settings. These find their way into the construction of fictional and conversational texts in a systematic fashion throughout the world:

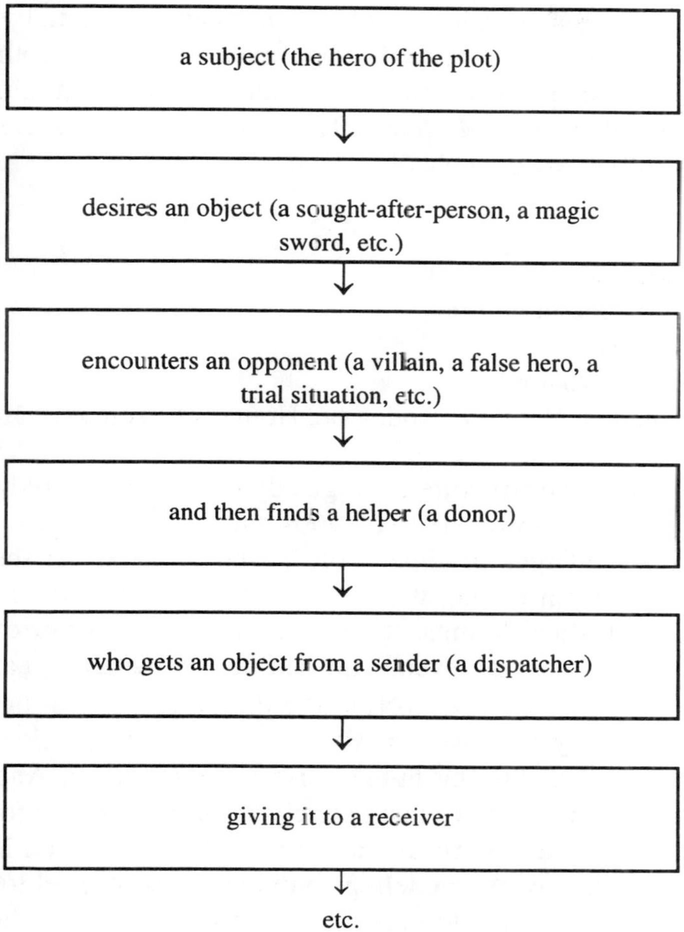

Figure 29. Greimas' narrative grammar

In order to explain the passage from these categories, which Greimas called *actants*, to actual narrative discourse, he posited a "generative trajectory", which maps the actants onto other constituents of a social interaction to produce the discourses that make up human narration and communication. An actant can be converted into various fundamental roles along a certain number of specified positions of its narrative trajectory. At the actual level of telling, one actant can be represented by several actors, and several actants by one and the same actor. In a mystery novel, for instance, the subject, or hero, may have several enemies, all of whom function actantially as an opponent. In a love story, a male lover may function as both object and sender. A simple example of how actantial theory might be applied to a novel such as *Madame Bovary* (1857) by Gustave Flaubert (1821-1880) goes somewhat like this:

subject = Emma
object = happiness
sender = romantic literature
receiver = Emma
helper = Léon, Rodolphe
opponent = Charles, Yonville, Rodolphe, Homais, Lheureux.

The study of narrative structure led logically to the serious study of myth within semiotics in the 1950s, 1960s and 1970s. The word *myth* derives from the Greek *mythos* 'word', 'speech', 'tale of the gods'. It is an early form of narrative in which the characters are gods, heroes, and mystical beings, the plot is about the origin of things or about dramatic human events, the setting is a metaphysical world juxtaposed against the real world, and the narrator is an unknown, perhaps metaphysical, source. Myths create a metaphysical knowledge system for explaining human origins and actions. And this system is the one to which we instinctively resort even today for imparting knowledge of the world initially to children. But even in contemporary adult life, mythic modeling continues to be utilized for various social purposes. Climatologists, for example, refer to the warming of the ocean surface off the western coast of South America that occurs every 4 to 12 years when upwelling of cold, nutrient-rich water does not occur as *El Niño,* 'the little one' in Spanish. This creation of a *character* to represent a climatological referent makes the referent much more understandable in human connotative terms. Although people do not think of *El Niño* literally as a person, they nonetheless find it convenient to blame "him", rather than some abstract process, for certain weather conditions. This mirrors, no doubt, how the original mythic characters were created–the difference being that the mythic characters of the past were believed to be real gods or mythical beings, not connotative extensions of events.

Because myth is a narrative, many attempts to understand it have focused on its composite structure. The most interesting interpretation for MST comes from the work of the French anthropologist Claude Lévi-Strauss (1908-), who saw myth as the original source for the parts of speech in language. Lévi-Strauss pointed out that certain clusters of relationships in myth conform to the systematic order of the language's structure: the first noun subjects were divine characters, the first noun objects were things and people in the world, upon which the gods acted; the actions of the gods thus constituted the first verbs. This same view was held two centuries earlier by the Italian philosopher Giambattista Vico, who also proposed the idea that mythic themes, characters, etc. led to the founding of the first civil institutions. In effect, the ancient myths were composite models of human ideas, emotions, and social and ethical behaviors. Not possessing the knowledge to understand or explain natural and human phenomena in scientific terms, the first humans ascribed them to awesome and frightful "gods" or "divine" creatures, thus producing humanity's first *archetypes* (literally, original *models*).

3.3.2 Stable vs. pliable models

Most myths, fictional works, and other kinds of recorded texts endure across time. Conversations, on the other hand, do not, unless they are recorded graphically or mechanically for some reason. This suggests two types of artificial modeling processes: *stable* and *pliable*. The main difference between discourse and, say, a fictional narrative, lies in the fact that discourse is a *modeling-in-progress*, or *pliable*, form of representation, whereas the written narrative is something that has become a *permanent* or *stable* form. The same distinction applies to natural modeling processes.

- *Natural Models:* A stable natural model has a form that is predictable within a certain range: i.e., the basic form can always be recognized in its many manifestations which are constrained by natural processes. For example, a wart is recognizable as a wart no matter how large it is in diameter. Nevertheless, the diameter of warts rarely supersedes a certain length. A pliable natural model is one that is adaptive to environment. For example, the type of song that certain species of birds learn permits considerable latitude. Any song will do as long as it has a few essential features. Because the memorization is not quite perfect and admits some pliability, the songs of many birds have developed regional dialects and thus serve as vehicles for a kind of "cultural" behavior.

- *Artificial Models:* A stable artificial model has an enduring form that allows it to be transmitted across space and time: e.g., a drawing, a word (that has gained currency), a narrative text recorded in some graphic form (pictographic, orthographic, etc.), and so on. Pliable artificial modeling is adaptive and responsive to the dynamics of reference and communication: e.g., discourse is a pliable form of modeling because it can be adjusted to reflect the changing needs of a communicative interaction. In a phrase; stable models are *fixed*; pliable models are *adaptive*.

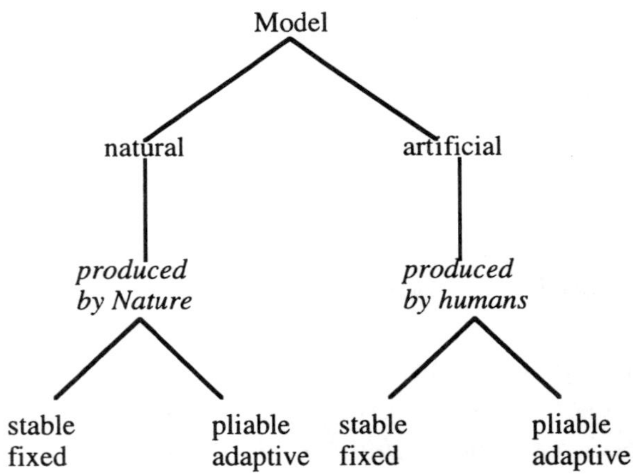

Figure 30. Types of models

Among the various semioticians who have studied the pliability of verbal discourse, the Moscow-born linguist and semiotician who carried out most of his work in the United States, Roman Jakobson (1896-1982), is perhaps the one who has provided the most important insights. Jakobson (1960) posited that discourse is hardly ever a *stable* information-transfer process, similar to the signal exchanges between animals to convey urges, needs, etc. Rather, he saw it as an adaptive, pliable form of verbal modeling that is constantly responsive to the conditions under which it is constructed.

3.4 Secondary cohesive modeling

Codes such as language, personal name-giving, musical notation, numeration, and the like, can also be extended morphologically and connotatively through secondary modeling. As an example, consider the musical code of classical Western music. This provides the tonal signifiers for making major and minor chords (chapter I, §1.2.2): a major chord is perceivable as distinct from a minor chord in the same key by virtue of a half tone difference in the middle tone of the chord:

- e.g., **C**-major, is made up of the sequence **C-E-G** on the piano keyboard;
- e.g., **C**-minor is made up of the sequence **C-E^b-G**, on the piano keyboard.

Now, either one of these triads can be extended morphologically as follows:

- Each chord can be given more "depth" or "sonority" by adding the octave higher tonic to the chord: **C-E-G-C** and **C-E^b-G-C**.
- Each triad can be played with a different combination of its constituent notes, producing the so-called "inversions" for different sonorous effects:

first inversion:	**E-G-C** and **E^b-G-C**
second inversion:	**G-C-E** and **G-C- E^b**

The interesting thing to note is that, in this case, morphological extensions produce (emotive) connotative effects concomitantly: the four-tone version of the chord sounds more sonorous and thus more fulfilling; the second inversion of a chord sounds "unresolved" and thus leaves one in suspension; and so on.

As we have mentioned various times, language is, by definition, a secondary cohesive modeling system providing humans with the resources for extending primary forms ad infinitum. As an exceedingly sophisticated modeling device, in the sense of von Uexküll's *Umweltlehre* (Lotman 1977), language was probably present in *Homo Habilis*. This ancestral member of our genus appeared, rather abruptly, about two million years ago. Language became "exapted" (Gould and Vrba 1982) in the species *Homo Sapiens* a mere 100,00 to 300,000 years ago in the form of speech. It took that long for the encoding abilities of *Homo Sapiens* to become fine-tuned with the species' corresponding decoding abilities.

Charles Morris (1964: 60) defined the language code as a *lansign-system*, a term that he applied not only to spoken and written languages but also to mathematics and symbolic logic, "and perhaps to the arts". His proposal to replace the word *language* with *lansign system* did not catch on widely with linguists or semioticians, but he was correct in observing that most linguists who have given the matter any thought at all would come to view their discipline as a semiotic study of the lansign-system, i.e., of language as a sign system. The one who followed Morris' suggestion more than anyone else was Louis Hjelmslev (Trabant 1981). In Hjelmslev's conception, natural languages were, *ipso facto*, natural lansign systems (Eco 1984: 14). But his program for the study of language has never been carried out for the simple reason that it is too idiosyncratic and highly complex (Sebeok 1985:13).

From a biosemiotic perspective, the language code can be defined as the cohesive system providing the modeling resources for converting what von Uexküll (1909) called "concrete living existence" into "active plans". Von Uexküll pointed out that the only way in which we can observe the mind is during the time in which it receives and works out impressions according to its activity or else in its products (signs, texts, etc.). The PMS enables children to acquire and compose a working knowledge of their world before they acquire facility with extensional verbal modeling. Interestingly, once the PMS has served this rudimentary function, it by no means disappears in adulthood and old age. In other words, the two modeling repertoires–the PMS and the SMS–remain intrinsically interwoven throughout the human life span, both complementing and supplementing one another. This reliance on two independent but subtly intertwined modeling strategies is what is unique about anthroposemiosis, rather than the mere language faculty in itself. As Bateson (1968: 614), has put it, the "decay of organs and skills under evolutionary replacement is a necessary and inevitable systemic phenomenon"; and this is why "if verbal language were in any sense an evolutionary replacement of communication by [non-verbal] means...we would expect the old...systems to have undergone conspicuous decay". But, as he continues, "Clearly they have not...Rather, the [non-verbal sign uses] have become richer and more complex, and [non-verbal communication] has blossomed side by side with the evolution of verbal language". The power of the language code lies both in the fact that it allows for both stable and pliable modeling (§3.3.3), and that its forms (words, phrases, sentences, conversations, etc.) can be extended morphologically and connotatively ad infinitum to encompass as many meanings as the human imagination deems necessary.

3.4.1 Name-giving codes

The *name-giving* codes that are found in cultures across the world constitute a perfect case-in-point of how an indexical cohesive code functions in social life. The study of *names* falls more properly under the branch of both semiotics and linguistics called *onomastics* (from Greek *onoma* 'name'). The phenomenon of name-giving in the human species is indeed a fascinating one. Across cultures, a neonate is not considered a full-fledged member of the culture until s/he is given a name. The act of naming a newborn infant is h/er first rite of passage in society. The name identifies the child as a separate individual with a unique personality. If a person is not given a name by h/er family, then the society in which s/he is born will step in to do so. A person taken into a family, by marriage, adoption, or for some other reason, is also typically assigned the family's name. From childhood on, individuals typically feel that their Self is somehow shaped by their name. In Inuit cultures, for instance, an individual is perceived to have a body, a soul, and a name; a person is not seen as complete without all three.

The name-giving code provides appropriate identifier signs for people. In Anglo-American culture, given (or first) names can indicate such things as: (1) a month or object (*May, June, Ruby, Daisy*); (2) popular contemporary personalities (*Elvis*, *Marilyn*); (3) classical mythical personages (*Diana, Jason*); or places (*Georgia*). Until the late Middle Ages, one personal name was generally sufficient as an identifier. Duplications, however, began to occur so often that additional differentiations became a necessity. Hence, *surnames* were assigned regularly to individuals (literally "names on top of names"). These were at first either indexical, in that they identified the individual in terms of place of h/er place of origin or parentage (descendancy), or descriptive, in that they identified the individual in terms of some personal or social feature (e.g., occupation). Thus, in England a person living near or at a place where apple trees grew might have been called "Mary who lives nearby where the apples grow", hence, *Mary Appleby.* Indeed, place surnames, such as *Woods*, *Moore*, *Church*, or *Hill* constitute a large number of English surnames. Descriptive surnames such as *Black, Short, Long,* etc. were created instead to highlight various personal or social characteristics. Descendant surnames, or names indicating parentage, were often constructed by prefixation–e.g., *Mac-*, *Mc-* in Scottish or Irish names or *Ap-* in Welsh names–or by suffixation–e.g., *-son* in English surnames and *-sen* in Scandinavian surnames (*Johnson* or *Jensen*, 'son of John', *Maryson* 'son of Mary', *Jakobsdottir*, 'daughter of Jacob'). Surnames reflecting medieval life and occupations–*Smith* being the foremost surname with equivalents in Spanish (*Ferrer*), German (*Schmidt*), and Hungarian (*Kovacs*), *Farmer, Carpenter, Tailor, Weaver,* etc.–also assumed an identifier function widely in the later medieval period.

Naming is not limited to human beings. In the animal world there are systems that have a comparable identifier function. These vary as to species, reproductive status, rank in a social hierarchy, momentary mood, and the like (Sebeok 1972a: 130). The best organized societies of vertebrates are distinguishable by a single trait (identifier) that is so overriding in its consequences that the other characteristics seem to flow from it. Wilson (1971: 402) draws a pivotal distinction between the impersonal (non-naming) societies formed by the insects, on the one hand, and the personal (naming) societies found in bird and mammal species, on the other. Each member of a naming society bears some particular relationship to every other member, and thereby comes to be known to all others as unique. Coupled with efforts to establish and maintain the requisite network of multifarious bonds is the development of an intimate form of communication, which necessarily involves the use of appropriate supportive signs. In studying birds, Thorpe (1967) has shown that when partners are absent, the remaining bird will use the sounds normally reserved for the partner, with the result that the partner will return as quickly as possible, as if called by name. Many other examples can be adduced from a study of various vertebrates, including canines, felines, primates and marine mammals (van Lawick-Goodall 1968; Rowell 1972). Whales apparently emit clicks that seem to have the same function of the so-called "signature-tunes" of birds. As Goffman (1963: 56) aptly puts it, in many species the notion of "uniqueness" implies the utilization of indicators, or "identity pegs".

3.4.2 Numeration codes

Extensional cohesive modeling can be seen in the ways in which numeration codes are created and then modified by human beings. Numeration codes provide the signifiers for representing the critical feature [quantity] in referents–[unit quantity], [two unit quantity], etc. All such codes were forged iconically. The earliest consisted simply of groups of straight lines, either vertical or horizontal, each line corresponding to the [unit quantity]: I (*one sheep*), II (*one ox* and *one other ox*), III (*one tree, another tree, and one other tree*), etc. As the need to represent larger [quantities] emerged, it became obvious that such simple iconic numeration strategies were inconvenient. As early as 3000 BC in Mesopotamia a special numeral was developed for the quantity [ten units]. The addition of this signifier made it possible to express the number 11 with two instead of eleven individual signifiers and the number 99 with eighteen instead of ninety-nine. Later, extra numerals for a number between 1 and 10, and additional numerals for numbers greater than 10 were fashioned by morphological extension.

In Babylonian cuneiform notation the numeral used for 1 was also used for 60 and for powers of 60; the value of the numeral was indicated by its context. The Egyptian hieroglyphic system used special numerals for 10, 100, 1000, and 10,000. The numeration code created by the ancient Romans had the merit of expressing all numbers from 1 to 1,000,000 with a total of seven numerals: I for 1, V for 5, X for 10, L for 50, C for 100, D for 500, and M for 1000. Roman numerals are composite forms which are read from left to right. The numerals representing the largest [quantities] are placed at the left; immediately to the right of those are the numerals representing the next largest [quantities], and so on. The [quantities] represented by numerals are usually added together to produce the value of the number: e.g., LX = 60, and MMCIII = 2103.

The numeration code in use in most parts of the contemporary world is the Hindu-Arabic one. This system was first developed by the Hindus in India in the third century BC, and was probably introduced into the Arab world about the seventh or eighth century AD. The first recorded use of the system in Europe was in the tenth century.

The key feature of the Arabic system was the regularized use of positional notation, whereby individual number signs assume different values according to their position in the written numeral. This was made practicable by the invention of a numeral for "zero". This sign makes it possible to differentiate between 11, 101, and 1001 without the use of additional numerals, and to represent any number in terms of ten symbols, the numerals from 1 to 9 plus 0. A further morphological extension of this system included the addition of exponential notation: e.g., the square exponent added to a number, 4^2, indicates a multiplication of the number by itself, 4×4; the cube exponent added to the same number, 4^3, indicates a multiplication of the number by itself three times, $4 \times 4 \times 4$; and so on.

3.5 Secondary connective modeling

Secondary connective modeling inheres in establishing linkages among already-forged metaforms. There are two types of secondary connective modeling processes. The first one can be called *layering* (chapter I, §1.6.2). Once the first "layer" of abstract metaforms in a language has been formed, on the basis of concrete source domains, then this layer itself becomes a new productive source domain for creating higher (= more abstract) layers of concepts. The linkages that result from layering can be called *meta-metaforms* (§1.6.2).

A simple graphic representation of this process is provided in figure 31 below:

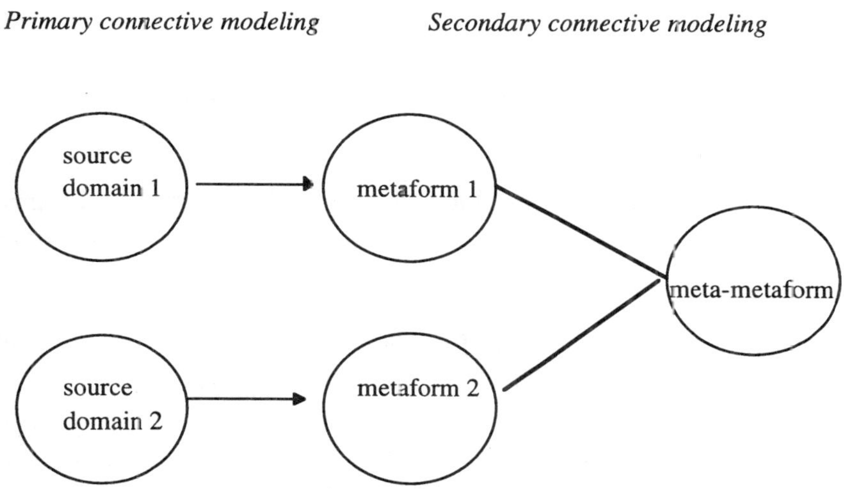

Figure 31. Secondary connective modeling

The other type of secondary connective modeling is called *cultural modeling*. This inheres in the association of various source domains with one target domain, producing an overall, or *culture-specific model*, of the target domain. We will look at both of these types of modeling processes below.

3.5.1 Meta-metaforms

As a practical example of how layering unfolds, consider once again the target domain of [thinking] (§1.6.2), which can be rendered, for instance, by source domain vehicles based on an [upward motion] image schema. This produces, of course, the metaform [thinking = upward motion], which underlies externalizations such as the following:

91. When did you think that *up*?
92. That problem comes *up* often, doesn't it?

The same target domain, [thinking], however, can also be rendered by source domain vehicles based on a [scanning motion], producing the metaform [thinking = scanning motion], which underlies externalizations such as the following:

93. Think *over* what you just said.
94. Think that problem *over* carefully before trying to solve it.

A linkage of these two produces the meta-metaform: [thinking = upward + scanning motions], as can be seen in the following externalizations:

95. That proposal *came out* of *nowhere*.
96. That notion *emerged* from the *domain* of psychology.

Expressions such as *come out of nowhere* and *emerge from the domain* are products of the meta-metaform [thinking = upward + scanning motions]. These portray thinking as both (1) something that emanates upward, (2) from an intellectual terrain. Here are some other externalizations of this meta-metaform:

97. This idea *came out* of the *depths* of my mind.
98. That idea simply *popped up* from the *nowhere*.
99. Think *over* your theory from a broader *perspective*.

These show clearly how complex abstract thinking unfolds. Phrases such as *come out, think up, pop up,* etc. elicit a mental image of upward movement, thus portraying the abstract referent as an object rising physically from a kind of mental terrain; *think over* evokes the image of scanning with the mind's eye. Now, a thought *coming out of nowhere* elicits a double image: a territory with a limitless scanning range (*nowhere*) and a thought *coming out* at some indeterminate point in this territory.

Sometimes meta-metaforms surface not only in such lexical externalizations, but also, in the form of grammatical dichotomies, such as the one between *in* and *on* in sentences such as the following:

100. I read it *in* the newspaper.
101. I saw that *in* the latest issue of *Time Magazine*.
102. It was written *on* a notepad, not *on* a slip of paper.

The use of one or the other preposition is due to the structural effect of two different meta-metaforms on the conceptual organization of these sentences. Reading words *in* something elicits a mental image of their location inside a container, which, in turn, entails the related image of removing their meaning from the container (the newspaper, the magazine, etc.). Indeed, this is why a logical follow-up utterance to "I read it *in* the newspaper" is "Well, what did you get *out of* it?" More specifically, the use of this preposition is motivated by an association of the [mind = a container] metaform–e.g., "Who *put* those thoughts *in* your mind?" "*Remove* that thought *from* your mind"; etc.–with the [thoughts = objects] metaform–e.g., "I can't quite *grasp* what you mean by that word"; "The *things* she said should be discarded"; etc.–producing the meta-metaform: [thoughts evoked by words = objects *in* containers]. The [containers] in this case are such things as newspapers and magazines. On the other hand, reading words *on* something elicits a mental image of words lying *on* a surface which, in turn, entails the related image of looking at the words. This is why the same type of follow-up above is a non sequitur in this case: i.e., it would be unusual to follow "It was written *on* a notepad" with "Well, what did you get *out of* it?" In this case, the use of the preposition *on* is motivated by an association of the [mind = a surface] metaform–e.g., "*Lay* your thoughts *out* for everyone to see?"–with the [thoughts = objects] metaform above, producing the meta-metaform: [thoughts evoked by words = objects *on* surfaces]. The [surfaces] in this case are such things as notepads and slips of paper.

A *meta-metaform* is, as such examples show, a secondary connective model that results when already-existing metaforms are connected to each other. The process of constructing the above [thinking

= upward + scanning motions] meta-metaform, for example, is shown graphically in figure 32 below:

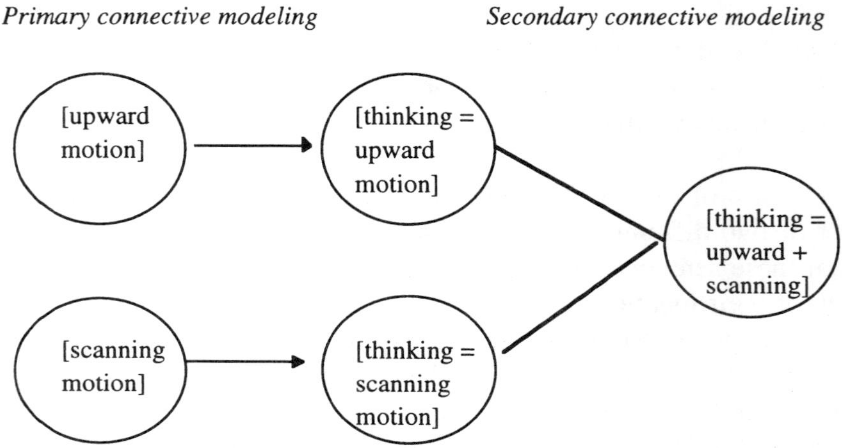

Figure 32. The meta-metaform [thinking = upward + scanning motions]

The layering of metaforms to produce higher abstractions (meta-metaforms) is an unconscious process. The higher the density of layering, the more abstract and, thus, more culture-specific, the concept (e.g., Dundes 1972; Kövecses 1986, 1988, 1990). Primary connective models like the [thinking = seeing] one (chapter I, §1.6.2) are relatively understandable across cultures: i.e., people from non-English-speaking cultures could easily figure out what the statements that instantiate this metaform mean if they were translated to them, because they connect a concrete source domain–e.g., [seeing]–to an abstraction–[thinking]–directly. Meta-metaforms, on the other hand, are more likely to be understood primarily in culture-specific ways, and are thus much harder to translate, because they connect already-existing metaforms to each other.

3.5.2 Cultural models

The other type of secondary connective modeling inheres in the association of various source domains to a single target domain, producing an overall, or *culture-specific,* model (Lakoff and Johnson

1980) of the target domain. Take, for example, a target domain such as [ideas], which is delivered by a large array of source domains. Here are a few of them in English:

[ideas = food]

103. What he said left a *bitter taste* in my mouth.
104. I cannot *digest* all that information..
105. He is a *voracious* reader.
106. We do not need to *spoon feed* our students.

[ideas = people]

107. Darwin is the *father* of modern biology.
108. Medieval ideas are *alive* and *well.*
109. Artificial Intelligence is still in its *infancy*.
110. She *breathed* new life into that idea.

[ideas = clothing/fashion]

111. That idea is not in *vogue* any longer.
112. New York has become a center for *avant garde* thinking.
113. Revolution is *out of style* these days.
114. Studying semiotics has become quite *chic.*
115. That idea is an old *hat.*

[ideas = buildings]

116. That is a *well-constructed* theory.
117. His views are on *solid ground.*
118. That theory needs *support*.
119. Their viewpoint *collapsed* under criticism.
120. She put together the *framework* of a theory.

[ideas = plants]

121. Her ideas have come to *fruition.*
122. That's a *budding* theory.
123. His views have contemporary *offshoots*.
124. That is a *branch* of mathematics.

[ideas = geometrical figures]

125. I don't see the *point* of your idea.
126. Your ideas are *tangential* to what I'm thinking.
127. Those ideas are logically *circular.*

[ideas = commodities]

128. He certainly knows how to *package* his ideas.
129. That idea just won't *sell.*
130. There's no *market* for that idea.
131. That's a *worthless* idea.

Now, a cultural model of [ideas] is simply the complex of all possible source domains to which it is associated–for example, [food], [buildings], [plants], [commodities], [geometry], and [seeing]:

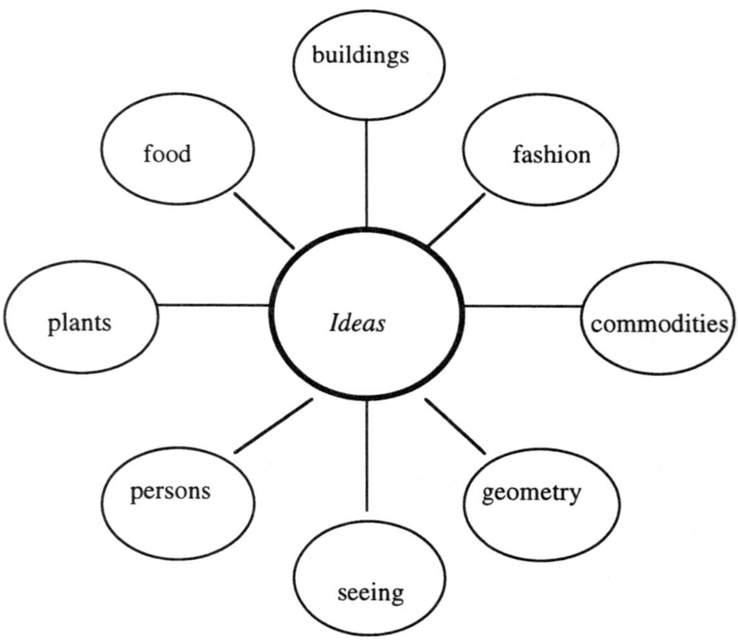

Figure 33. Cultural model of *ideas*

There are many more source domains for conveying the concept of [ideas] in English. The point to be made here is that the specific configuration of source domains produces an overall cultural model of a concept. Cultural groupthink is built on such models, since these coalesce into a system of abstract meaning that holds together the entire network of associated meanings in the culture.

Chapter IV
Tertiary modeling

In fact, words are well adapted for description and the arousing of emotion, but for many kinds of precise thought other symbols are much better.

J. B. S. Haldane (1892–1964)

4. Introductory remarks

As defined in the opening chapter (§1.2), the TMS is the system that undergirds highly abstract, symbol-based modeling. Like the SMS it is an extensional system, allowing for the further expansion of forms to encompass larger and more abstract domains of reference.

In this chapter we will look at the nature of symbolicity from the biosemiotic perspective and at the kind of modeling phenomena that the TMS permits. In this chapter we will be using two terms *symbolicity* and *symbolism.* The former is used to indicate the production and use of symbols in representation; the latter to "symbolic meaning" in general. The human TMS is the ability that emerged to make *Homo Sapiens* doubly sapient–hence the designation *Homo Sapiens Sapiens*. Indeed, the distinguishing characteristic of the human species has always been its remarkable ability to represent the world in the form of complex symbols. This ability is the reason why, over time, humanity has come to be regulated not by force of natural selection, but by "force of symbols", i.e., by the accumulation of the meanings that previous generations have encoded in the form of symbols and passed on in cultural settings. But, as we shall see in this chapter, tertiary modeling is not unique to anthroposemiosis; it can be found in the zoosemiotic and phytosemiotic realms as well. The objective of SA is, again, to document all manifestations of tertiary modeling in all semiosic spheres.

4.1 The tertiary modeling system

Tertiary modeling implies, above all else, the ability to extend forms to stand for abstract referents freely, without any apparent sensory connection between the form and the referent. It also entails the ability to utilize forms creatively and resourcefully. Tertiary modeling is especially prevalent in the human realm. The presence of symbolic forms in representational systems across the world is evidence that human consciousness is not only attentive to physical patterns (color, shape, size, etc.), resulting in iconic representational activities, and cause and effect patterns (contingent on time and space constraints), resulting in indexical representational activities, but also to pattern in itself. The end-product of this form of attentiveness is symbolic representation.

But creative tertiary modeling is not unique to anthroposemiosis. Gulls, for instance, apparently have the ability to modify their threat displays during courtship in a creative manner. Male spiders with well-developed eyes often utilize bright color patterns creatively to avoid having the female treat it as food. These examples abound, remarkably, in the zoosemiotic domain.

In the human realm, nowhere has the TMS borne more remarkable fruits than in the area of mathematical and scientific representation. The science of geometry, for instance, has helped human beings solve engineering dilemmas since ancient times. Here is a typical example of how symbolicity in this domain of representational activity has allowed humans to plan activities beforehand in an ingenious manner.

Suppose a tunnel must be opened up through the middle of a large boulder. Obviously, the length of the tunnel cannot be measured directly. However, the availability of the so-called *Pythagorean Theorem* allows us to devise an ingenious strategy for measuring it indirectly:

- A point **A** on one side of the boulder and another point **B** on the other are chosen such that both points remain visible from a point **C** to the right.
- **C** chosen so that angle **ACB** is a right angle (90^0).
- Then, by aligning **A** with **A´** (the entry point to the boulder on one side) and **B** with **B´** (the exit point from the boulder on the other side) the required length can be determined.

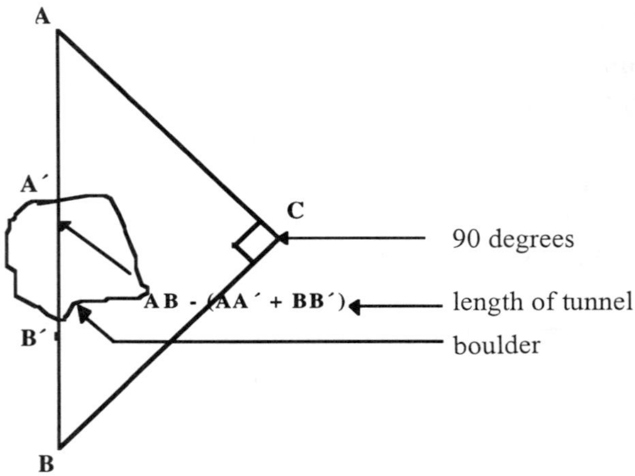

Figure 34. Engineering diagram suggested by the Pythagorean Theorem

It is, clearly, a straightforward task to measure **AC** and **BC**, since these lengths were chosen on purpose to permit their measurability. Now, the Pythagorean Theorem tells us that $\mathbf{AB}^2 = \mathbf{AC}^2 + \mathbf{BC}^2$. This general formula allows us to determine the length of **AB** easily because the lengths **AC** and **BC** can be determined by simple measurement. By subtracting the distances **AA´** and **BB´** from **AB**, **AB** - (**AA´** + **BB´**), the length of the tunnel can then be ascertained.

This simple example illustrates how the TMS allows humans to produce models of referents that can be manipulated and tested totally within the mind until they incorporate the array of features required to solve some real-world problem. A tertiary model is an idealization in the form of a referent, and can thus be applied to many kinds of problems that entail the same type of form.

4.1.1 Symbolicity

The debate on the origin of symbolicity in the human species goes back to the ancient Greeks. The philosopher Plato (427?-347? BC), for instance, viewed symbolic representation and especially language as separate from sensory imitation (iconicity). In Western culture, the one who entrenched this view even more deeply was the French philosopher René Descartes (1596-1650), who claimed to show that nonverbal forms of thought were without logic, and so could not be studied scientifically; whereas symbolic forms (verbal and mathematical) were inherently logical and thus the basis of human thinking. However, as we have emphasized throughout this book, such views ignore the fact that even the most abstract forms of representation, such as the ones used in mathematics, do not originate as purely symbolic. The diagram devised above in §4.1 is, in effect, an iconic composite form which was drawn up to represent in visual outline (and in compressed form) the actual physical scenario in question. So, too, the first notion that the sides of right-angled triangles were related to each other in some systematic way did not crystallize in someone's imagination *ex nihilo*. Rather, it took shape after repeated measurements of the three sides of right-angled triangular figures. Subsequently, perhaps by drawing squares on diagrams of right-angled triangles, someone must have noticed that the area of the square on the two sides of the triangle invariably added up to the area of the square on its hypotenuse (i.e., Area 1 = Area 2 + Area 3):

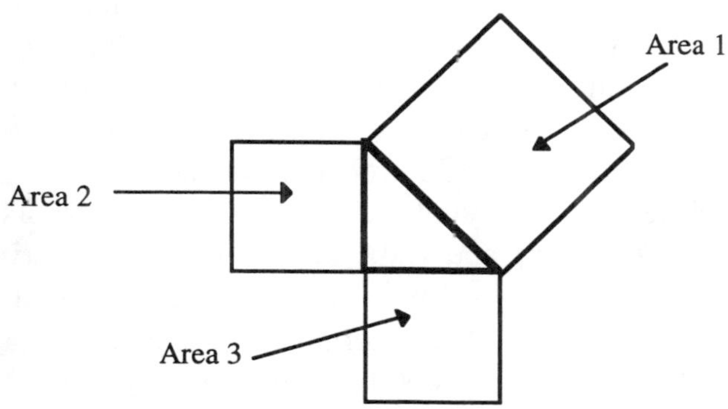

Figure 35. Visual demonstration of the Pythagorean Theorem

Eventually, someone "proved" this, paving the way for the establishment of the Pythagorean Theorem. Any theory in mathematics and in science is, in effect, the extension of a "diagram model" of something. As the ancient Greek geometers themselves emphasized, such models permit representations of various aspects of physical reality which, in turn, allow for an intellectual experimentation with that very reality through the use of acquired notions (such as the Pythagorean Theorem). The results of this experimentation can then be redirected to the real world to see what they yield (as we did above). In effect, diagrams allow scientists to visualize something with the mental eye that is unseeable with the physical eye. In the engineering problem above, the diagram allowed us to measure something with our mental measuring tape that could not be measured with a physical one. In line with the principle of extensionality, theorems such as the Pythagorean one, are really no more than tertiary extensions of iconic modes of thinking and representing.

The creation of symbols through extensional modeling has had truly remarkable consequences on the evolution of humanity. This is why the great twentieth-century philosopher Ernst Cassirer characterized the human being as a *symbolic animal*. Consider numeration codes again (chapter III, §3.4.2). As we saw, number signs started out as iconic forms: I = [one unit], II = [two units], III = [three units], etc. Such forms were stylized later, resulting in numerals which, in positional arrangements, could represent any number. The further extension of positional numerals, such as the Hindu-Arabic ones, led eventually to the generalization of number representation itself by which symbols, such as letters of an alphabet, could be used represent numbers or members of a specified set of numbers and related to each other by operations that hold for all numbers in the set.

The history of such tertiary representational activities, known as *algebraic*, began in ancient Egypt and Babylon. Greek mathematicians continued the traditions of these two societies, but at a much more sophisticated level. However, the emergence of algebra as a functional mode of mathematical representation for solving practical problems had to await the ninth century, when the Arab mathematician Al-Khwarizmi (780?-850? AD) wrote one of the first scientific treatments of algebra. By the end of the century, the basic laws of algebra were established, and by medieval times Islamic and Persian mathematicians had worked out the basic theory of equations. Algebraic symbols were introduced and standardized in the sixteenth century. In the subsequent century, René Descartes invented analytic geometry, a system for solving geometric problems algebraically.

Algebra entered its modern phase around 1800, when the attention of mathematicians shifted away from solving equations to studying the structure of abstract mathematical notions.

The symbols of the algebraic code include numerals, letters, and signs indicating arithmetic operations. Letters can represent either constants or variables. Grouping symbols such as parentheses ensure that the language of algebra is clearly read, by showing the order in which operations are to be performed. But when looked at closely, algebra is really no more than "arithmetic with letters". Hindu-Arabic numerals are extensions of primary forms; algebraic symbols are further extensions of these:

Primary modeling		*Secondary modeling*		*Tertiary modeling*
iconic numerical forms	→	extension of the forms	→	generalization of the numeral forms
I = [one unit], II = [two units], etc.		1 = [one unit], 11 = [eleven units], etc.		x = [any number], x^n = [any number multiplied by a given number of times], etc.

Figure 36. Extensional modeling of number

Artificial symbolic codes, such as the algebraic one, are exclusively products of the human TMS. However, more rudimentary manifestations of natural symbolicity are common throughout Nature (Pitts and McCulloch 1947; Haldane 1955: 387; Sebeok 1973a: 196; Jacob 1974). A rhesus monkey, for instance, shows fear by carrying its tail stiffly out behind; baboons convey fear by carrying a vertical tail. However, the converse is not necessarily true: "a mother of a young infant [baboon] may hold her tail vertical not in fear but to help her infant balance on her back; and the tail may also be held vertical while its owner is being groomed in the tail region" (Rowell 1972: 87).

This is, clearly, symbolic behavior given that tail orientation stands in an indirect fashion for an emotion. Consider, further, the behavior of the insects of the carnivorous family *Empididae*. In a species of dipterans of this family, the male offers the female an empty balloon prior to copulation (Huxley 1966). The evolutionary origin of this seemingly bizarre gesture has been unraveled by biologists. But the fact remains that the gift of an empty balloon is a wholly symbolic act, designed simply to reduce the probability that the male himself will fall prey to his female partner.

Examples such as these abound in the zoosemiotic realm. Indeed, the cross-species study of symbolicity suggests that the evolution of symbols in various species is an outcome of evolutionary processes. The difference in the human species is the diversity and range of symbolicity, and the fact that it is fully witting, rather than purely osmotic and instinctiual.

Human symbols have a general and variable applicability. Take, for example, the statement 2 + 2 in mathematics. As a symbolic statement, it will, by its very nature, be applicable in more than one way (i.e. to ore than one situation). Usually, it is interpreted as having the meaning 2 + 2 = 4. A "real-world" model to show the validity and workability of this interpretation is a container calibrated into four 4 units into which two buckets of water, each one calibrated into 2 equal units, can be poured. In terms of this model, it is easy to see that by pouring the two buckets consecutively into the container, the latter will become full to its highest calibration level 4 (see figure 37 below).

More specifically, the statement 2 + 2 = 4 is valid (and thus has specific applications to some real-world model) within a so-called *modulo 10* number system, i.e., a system with ten elements (digits)– 0, 1, 2, 3, 4, 5, 6, 7, 8, 9. However, if we modify the container structurally by adding a drainage spout to it, as shown below in figure 38, then it can be seen that after the first bucket is poured into it, only 1 unit of its water content will remain in the container, because the other unit would simply trickle out of the container through the spout. Obviously, no units of water contained in the second bucket can be poured into the container, since these too would spill out of the container through the spout. In summary, the container equipped with the drainage spout (as shown in figure 38) cannot be filled beyond the calibration level 1. So, as applied to this new model, the statement 2 + 2 is now seen to equal 1: 2 + 2 = 1.

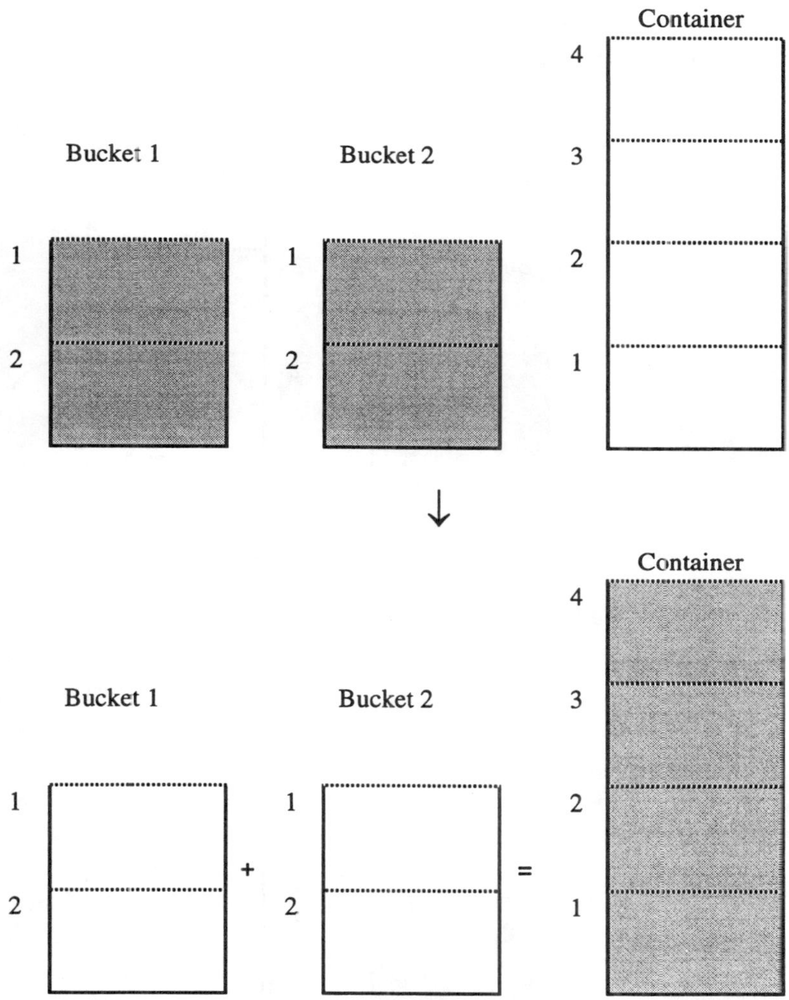

Figure 37. Container model of 2 + 2 = 4

The gist of the above exemplification is that when 2 + 2 is applied to a modified model a different result is obtained. The statement 2 + 2 = 1, incidentally, is valid within a so-called *modulo 3* number system, i.e., a system with just three elements–0, 1, and 2. The completed addition and multiplication tables of modulo 3 integers are shown in table 7 below. This examplehighlights the difference between *representation* and *application.* Models are, in effect, *representamina* (as Peirce called them). These may or may not have any *application* to the real world. When they do the models can be said to be *reifiable*. In this sense *application* is *reification*.

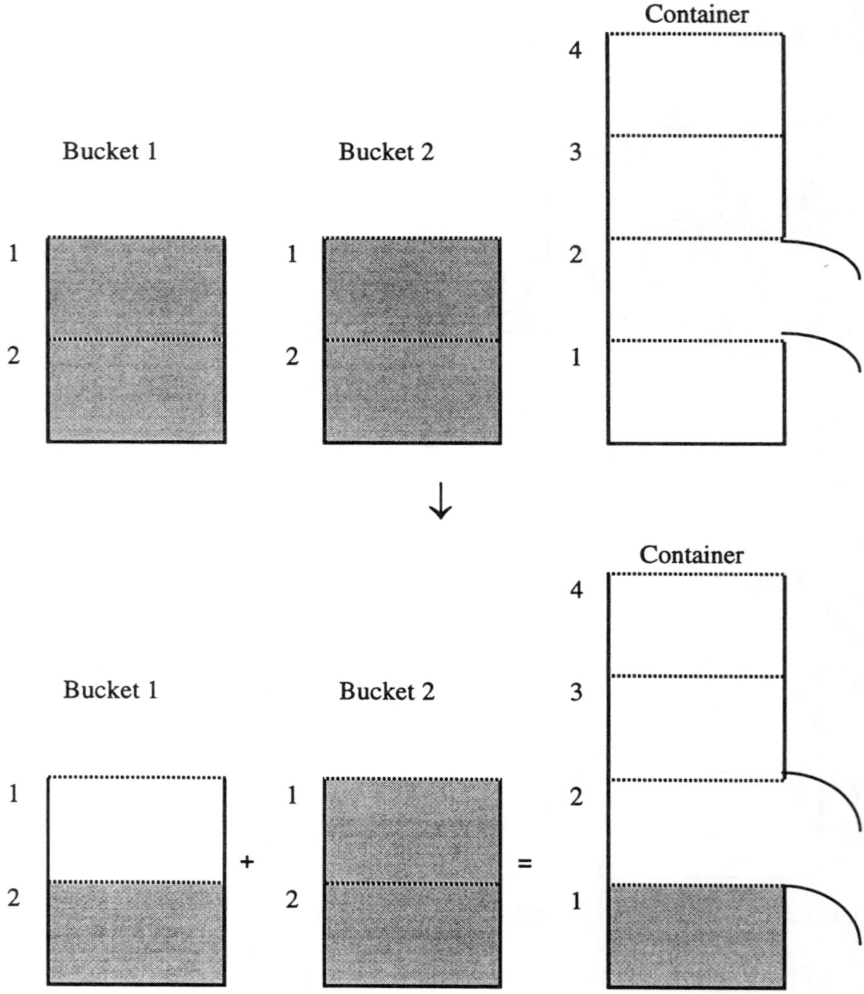

Figure 38. Container model of 2 + 2 = 1

Table 7. Modulo 3 arithmetic

+	0	1	2
0	0	1	2
1	1	2	0
2	2	0	1

×	0	1	2
0	0	0	0
1	0	1	2
2	0	2	1

As can be seen, in modulo 3 arithmetic, the statement 2 + 2 = 1 is, indeed, the correct one. Such multi-interpretations and multi-applications of the same forms constitute *the* salient characteristic of human symbolicity.

4.1.2 Culture

The signifying order (chapter I, §1.6.3) of a culture can be defined, *tout court*, as a cohesive TMS, i.e., as an interconnected system of signs, texts, codes, and connective forms. Perhaps the most fascinating thing about culture is that the symbolic models that it allows people to produce have a "rebounding effect", so to speak: i.e., they induce worldview, *mediating* knowledge of the world. A symbolic form of any kind (singularized, composite, etc.) specifies what is to be known and memorized from the infinite variety of things that are in the world. Although we create new forms all the time to help us gain new knowledge and modify previous knowledge, by and large, we literally let our cultural symbols "do the thinking" for us most of the time. As social beings, we are born into an already-fixed macro-code of forms that will largely determine how we will come to view the world around us. Only if, hypothetically, all our knowledge were somehow erased from the face of the earth, would we need to rely once again on our instinctive semiosic tendencies to model and represent the world all over again.

This discussion is meant to emphasize how crucial the role played by culture is in human social and cognitive life. But this does not mean, as the German philosopher Friedrich Nietzsche (1844-1900) despondently believed, that human models are no more than tokens of artifice and, therefore, that nothing outside of the human imagination is real. After all, the theory of aerodynamics can be used rather successfully to fly planes; theories about cell structure and function are used by doctors all the time to cure various diseases; and so on. The ability to represent the world with a vast array of symbolic forms (singularized, composite, cohesive, and connective) allows humans to experiment with the real word, and even to alter it to fit specific needs.

One of the more widely-known explications of the relation between culture and Self-experiences is the one put forward by the philosopher Karl Popper (1902-1994). Popper classified human experience into three "Worlds". "World 1" is the experience of physical objects and states as processed by neuronal synapses–electrical impulses between brain cells–transmitting messages along nerve paths that cause muscles to contract or limbs to move, and sensory systems to respond to perceptual input. "World 2" is the domain of subjective experiences. This is the level at which the concept of Self emerges, whereby the individual person develops the ability to differentiate h/erself from the beings, objects, and events of the outside world. It is at this level that Self-experiences such as perception, planning, remembering, dreaming, and imagining unfold. "World 3" is the domain of culture-specific knowledge that constrains Self-experiences. This is the level at which the signifying order enters into the individual's life to provide the categories of knowledge and representation that s/he will come to use routinely and habitually for h/er daily life.

4.2 Tertiary singularized modeling

Tertiary singularized models are, in effect, single symbols. According to the extensionality principle (chapter I, §1.2.1), the tendency in human representation is to create singularized forms iconically first and then to extend them subsequently to encompass abstract referents. Consider the word *flow* one more time (see also chapter II, §2.1.2, chapter III, §3.2). We argued that, in all likelihood, this word was coined iconically (probably as far back as the prehistoric Indo-European period) to simulate the sound made by the movement of water or of fluid pouring forth. But the modeling history of this form does not end there. As we saw, the distinctive features of its iconic denotatum–[continuous], [smooth], [movement]–can be extended freely to encompass other referential domains (connotata), of varying degrees of abstraction:

flowing traffic	=	[continuous]
flowing words	=	[smooth] and [continuous]
a *flow* of ideas	=	[continuous]

Now, this very same signifier can be used for whatever representational need that may arise. For example, it has been used to designate the sequential diagrams employed to show the stepwise procedures used in performing a task. These are called *flowcharts* because they imply the original iconic features associated with *flow*– namely, [continuous], [smooth], [movement]. It has also been used, in modified form as the word *fluency*, to refer to [facility in the use of a language]–an ability that also implies [continuous], [smooth], [movement]. The modeling history of *flow* is shown in figure 39 below:

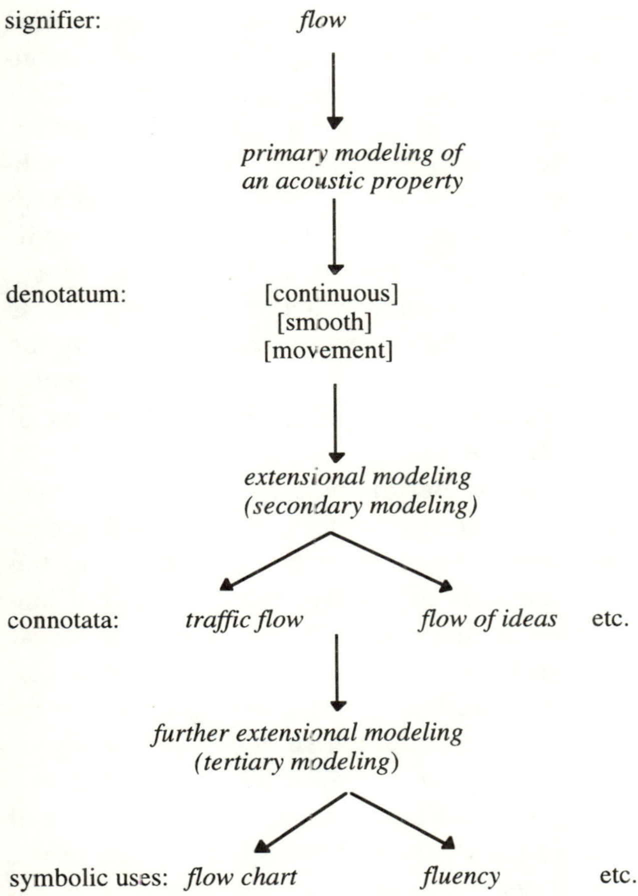

Figure 39. Modeling history of *flow*

4.2.1 Tertiary verbal singularized modeling

Although symbolic forms (like the *flowchart* concept) stand for their referents in arbitrary, conventionalized ways, they are nevertheless extensions of more fundamental iconic features; i.e., the congener of the word *flowchart*, for example, would have had to know the features of the original signified of the English word *flow* in order to think up such a use of this word.

An area of representation that can be used to show how singularized symbolic forms have evolved out of more sensory forms of representation is *color terminology*. Physicists claim that there are potentially 8 million hue gradations that the human visual-perceptual system is capable of distinguishing. This capacity can be demonstrated by putting a finger at any point on the light spectrum–which consists of a continuous gradation of hue from one end to the other. A minuscule, yet real, difference in gradation in the colors immediately adjacent to the finger at either side would be easily picked up by the eye. Yet, a speaker of English describing the spectrum will probably list the two gradations as falling under the same category: i.e., as *purple, blue, green, yellow, orange, red*, etc. This is because the speaker has been conditioned by these very terms to classify the content of the spectrum in specific ways. There is nothing inherently "natural" about the speaker's classificatory decision; it is a reflex of English vocabulary, not of Nature.

In 1969, the psycholinguists Berlin and Kay argued, however, that differences in color terms are only superficial matters that conceal general underlying principles of color perception. Using the judgments of the native speakers of twenty widely-divergent languages, Berlin and Kay came to the conclusion that there were "focal points" in basic (single-term) color systems which clustered in certain predictable ways. They identified eleven universal colors, or focal points, which corresponded to the English words *red, pink, orange, yellow, brown, green, blue, purple, black, white,* and *gray*. Not all the languages they investigated had separate words for each of these colors, but there emerged a pattern that suggested to them the existence of a fixed way of perceiving color across cultures. If a language had two colors, then the focal points were equivalents of English *black* and *white*. If it had three color terms, then the third one corresponded to *red.* A four-term system had either *yellow* or *green*; while a five-term system had both of these. A six-term system included *blue*; a seven-term system had *brown*. Finally, *purple, pink, orange,* and *gray* were found to occur in any combination in languages that had the previous focal points. Berlin and Kay found that languages with, say, a four-term system consisting of *black, white, red*, and *brown* did not exist. The Berlin-Kay model of universal color coding is shown graphically in figure 40 below:

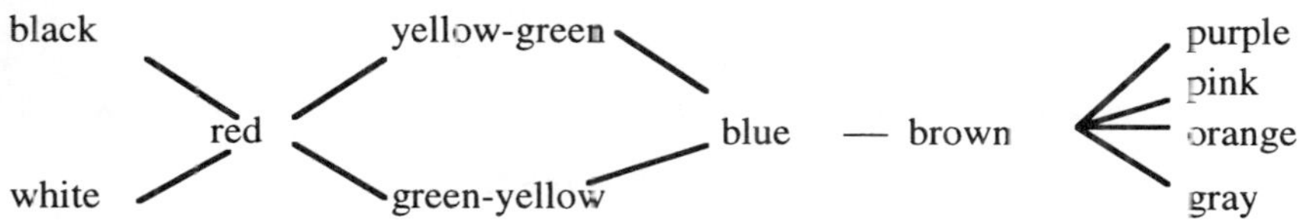

Figure 40. Berlin-Kay model of color coding

The intriguing implications of this research were pursued vigorously in the 1970s by many psychologists. Eleanor Rosch (e.g., 1975a, 1975b), for instance, demonstrated that the Dani people of West Irian, who have a two-color system, were able to discriminate easily eight focal points. Using a recognition-memory experiment, Rosch found that the Dani recognized focal colors better than non-focal ones. She also found that they learned new colors more easily when the color names were paired with focal colors. Such findings suggested to Rosch that languages provided a guide to the interpretation of color, but that they did not affect its perception in any way.

But, semiotically speaking, this line of research fails to account for the fact that speakers of different languages are predisposed to see different color categories on the very same spectrum through the repertoire of specific signifiers they have learned, not focal points. These may indeed correspond to certain English-language terms, in a certain sequence (as specified above). But the fact remains that a color signified is just what a color signifier says it is. Speakers of Shona, an indigenous African language, for instance, use four terms *cipswuka, citema, cicena,* and *cipswuka* (again), and speakers of Bassa, a language of Liberia, just two, *hui* and *ziza,* to refer to the gradations of hue on the spectrum. The relative proportional widths of the gradations that these color categories represent vis-à-vis the English categories are shown graphically in figure 41 below:

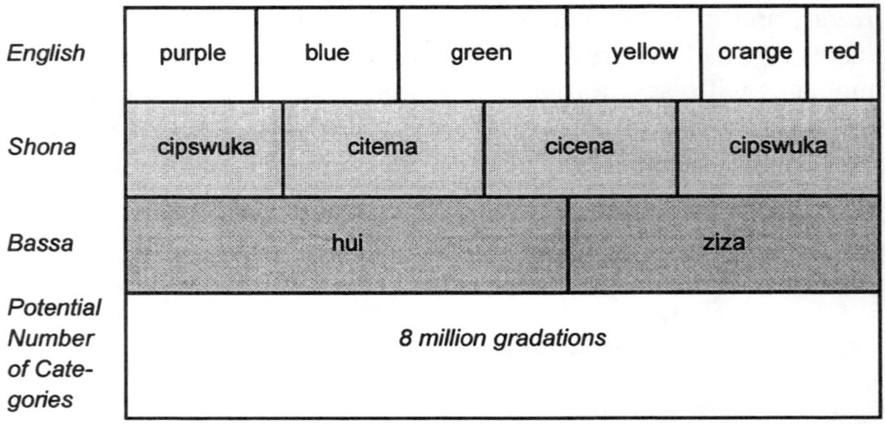

Figure 41. English, Shona, and Bassa color categories

So, when an English speaker refers to, say, a ball as *blue,* a Shona speaker might refer to it as either *cipswuka* or *citema,* and a Bassa speaker as *hui.* As the color researchers aptly showed, this does not stop an English speaker from relating h/er categories to those of the other two languages, or vice versa. We have, in fact, done this ourselves with the above chart. Indeed, the color terms one has learned in cultural context in no way preclude the ability to perceive the color categories encoded by the languages of other cultures. This is what a learner of a foreign language ends up doing when s/he studies the new system of color representation: i.e., s/he learns how to reclassify the content of the spectrum in terms of new signifiers. Moreover, in all languages there exist resources for referring to more specific gradations of color should the situation require it. In English the words *crimson, scarlet, vermilion,* for instance, make it possible to refer to gradations of *red.* But these are still felt by speakers to be *subcategories* of red, not distinct color categories on their own. Using a small set of signifiers to refer to color gradations is a practical representational strategy; otherwise, millions of signifiers would need to be invented to classify the spectrum in terms of all the possible discriminations that can be made.

The archeological record strongly suggests that sensory and emotional meanings may have been the source for most of the color terms created throughout the world. Take, for instance, the English word *blue*. Its denotatum is, in effect, the hue perceived in a clear daytime sky. This [sky] signified is, clearly, the result of a visual sense inference. The word *blue* is of Indo-European origin and, in all likelihood, referred to the sky or the sea, or to some other light-colored phenomenon. As the anthropologist Roger Wescott (1980) has amply documented, color vocabularies seem to have originated from the specific need to name such phenomena as the sky. In Hittite, for instance, words for colors initially designated plant and tree names such as *poplar*, *elm*, *cherry*, *oak*, etc.; in Hebrew, the name of the first man, *Adam*, meant *red* and *alive*, and still today, in languages of the Slavic family *red* signifies *living*, *beautiful*.

Now, through connotative (secondary) extensional modeling *blue* can be extended in English-speaking cultures to encompass a host of abstract referents that imply the [sky] signified connotatively:

[sadness]
I'm feeling *blue* today.

[murky]
The air was *blue* with oaths.

[unexpected]
That came out of the *blue*.

The connotatum [sadness] implies that the [sky] is absent ([-sky]), leading to a "downcast" mood, which is felt to be typical on cloudy days; the [murky] connotatum derives, probably, from the fact that the same type of hue is perceived in sea water, which often appears to be murky; and the [unexpected] connotatum is an extension of the fact that something falling from the sky is, of course, unexpected and sudden.

Now, by further extension, the [sadness] connotatum has led to the naming of a certain style of popular music as the *Blues*. This is music characterized by minor-key harmonies, a typically slow tempo, and melancholy words. Blues music is based on a scale in which the third, fifth, and seventh notes are freely "bent" or flattened (called *blue* notes) in comparison with the standard major scale, imparting a sad mood. Blues lyrics tend to deal with the hardships of life and the vicissitudes of love.

The modeling source of the *Blues* is shown graphically in figure 42 below:

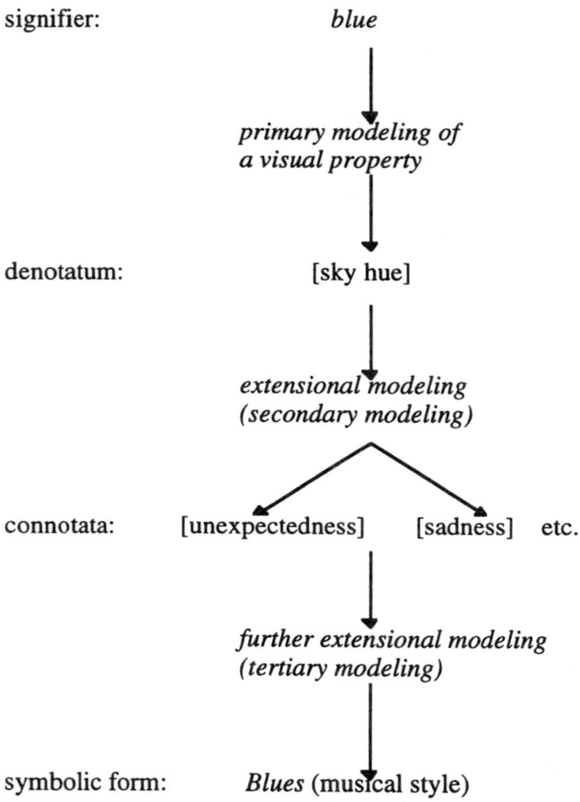

Figure 42. Modeling history of *Blues*

This type of tertiary modeling can be seen in the modeling histories of most of the color terms: e.g., one of the connotata of *red* is [blood], a meaning that has been extended further to designate the kinds of things that the *Red Cross* carries out.

In sum, the modeling history of any color term shows that the initial meaning of the term is linked to some sense inference. Once the term enters the language, it is extended to produce connotata. Finally, the connotata can be extended further to produce symbolic meanings and uses of the term. This three-stage modeling process, as applied to *blue*, is represented graphically in figure 43 below:

Primary modeling		*Secondary modeling*		*Tertiary modeling*
sense-inference produces a denotatum	→	extension of the sense-inference produces connotata, such as:	→	further extension of a connotatum (e.g., [sadness]) produces a symbolic form
blue = [sky hue]		*blue* = [sadness]		*Blues* = musical style

Figure 43. Extensional modeling history of the *Blues* concept

4.2.2 Tertiary nonverbal singularized modeling

Tertiary singularized modeling is found, of course, in the nonverbal domain of representation as well. Visual singularized forms, for instance, are often interpreted symbolically. Take, for instance, the following triangle:

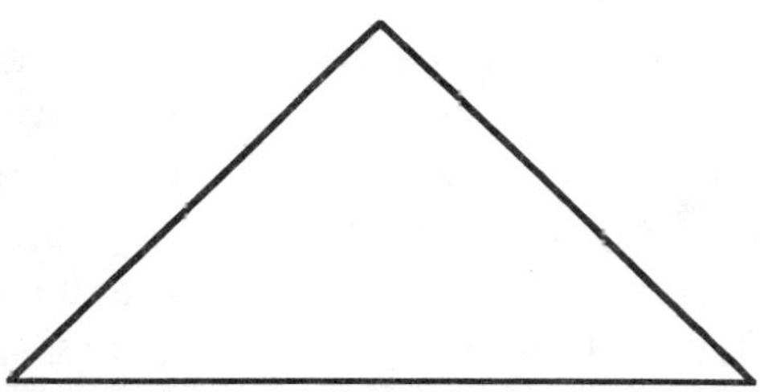

Figure 44. Equilateral triangle

This is an equilateral triangle, considered, in Western culture, to be the "ideal" or "prototypical" form of the triangle figure. Different triangular shapes produce different "aesthetic" effects. For example an obtuse triangle–a triangle that has one angle larger than 90^0–seems "slightly off" to the Western eye:

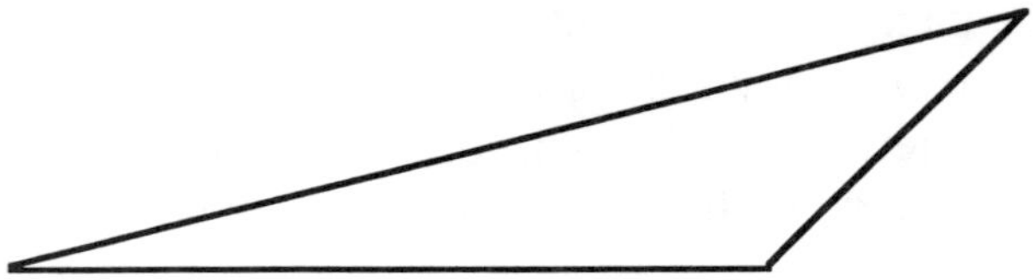

Figure 45. Obtuse triangle

This "slightly-off" effect spills over into the verbal domain. This is why we say the following things:

obtuse student	=	[unintelligent]
obtuse pain	=	[dull]
obtuse idea	=	[dull], [uninteresting]
obtuse style	=	[dull], [uninteresting]

The perception of figures as ideal or not is, clearly, a product of tertiary modeling. Since antiquity many philosophers, artists, and mathematicians have been intrigued by ideal figures. One such figure is called the *golden section*. This is a figure constructed in such a way so that the ratio of the length of the longer line segment to the length of the entire line is equal to the ratio of the length of the shorter line segment to the length of the longer line segment. This ratio, which has the numerical value 0.618.... is widely perceived as producing figures that exhibit special beauty.

Tertiary visual symbols abound in all kinds of cultural settings, bearing great meaning to the denizens of specific societies. Flags, flowers, emblems, heraldic symbols, totemic symbols, military symbols, etc. bear great significance in cultures throughout the world. Flowers, for instance, have been adopted by noble families (and emblazoned on their coats of arms), by countries, by cities, and, in Canada and the US by various individual states as official emblems. The golden chrysanthemum, native to eastern Asia, is the symbol of Japan and the imperial family. In France the fleur-de-lis was adopted as the royal emblem by King Louis VI in 1108. In England, the house of Lancaster, whose symbolic emblem was a red rose, and the house of York, whose emblem was a white rose, battled for the crown in the *Wars of the Roses* (1455-1458).

4.3 Tertiary composite modeling

A tertiary composite model is, in effect, a text that is imbued with a high degree of symbolicity. As we have seen, scientific texts are, by and large, tertiary composite forms. A mathematical equation for instance, is composed of symbols, known as variables or indeterminates, commonly denoted by letters or other symbols: e.g., $x^2 + y - 4 = 9$; $y = \sin x + x$; $3y = \log x$; etc.. In order to grasp the meaning of the referents of equations, it is necessary to understand how they model them. For example, the classic Pythagorean equation, $\mathbf{A}^2 + \mathbf{B}^2 = \mathbf{C}^2$ is a model of a right-angled triangle with sides **A**, **B**, and **C**, of which **C** is the hypotenuse:

$\mathbf{A}^2 = \mathbf{B}^2 + \mathbf{C}^2$

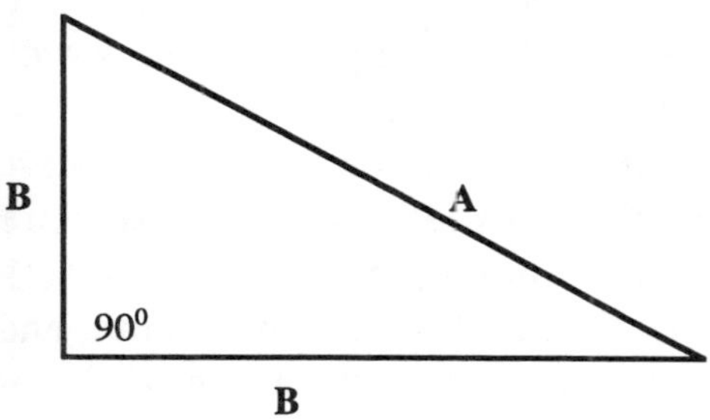

Figure 46. Right-angled triangle

However, as we discussed above (§4.1.1), tertiary forms such as this one, result from extensional processes. Many early civilizations considered the Pythagorean equation true, not because it could be proved by deductive logic, but because it agreed with their observations in practical situations. Its formalization and proof as a theorem came much later.

4.3.1 Tertiary verbal composite modeling

Verbal texts can be factual or fictional, as discussed in the previous chapter (§3.3.1). Fictional texts (novels, films, etc.) are interpreted as providing various psychological, sociological, or metaphysical meanings. Although we read a fictional novel or watch a fictional movie knowing that its plot, characters, and setting are fictitious, we still see it as modeling some real-life happening. This is because the entire narrative is interpreted as a single symbol (= *symbolism*).

Fictional narratives in prose were composed in the ancient world, and to these the term *novel* is sometimes applied. But the novel had not emerged as an autonomous narrative art form until the Middle Ages. Actually, many scholars regard the eleventh century *Tale of Genji*, by the Japanese Baroness Murasaki Shikibu (978?-1026?), as the first true novel, since it depicts the amorous adventures of the fictional Prince Genji and the staid lives of his descendants. Fiction can be said to start in the West with the long verse tale, the prose romance, and the Old French *fabliau* in the medieval period, culminating with Giovanni Boccaccio's (1313-1375) *Decameron*. Advances were made in Spain during the sixteenth century with the so-called *picaresque* novel, in which the protagonist is typically a vagabond who goes through a series of exciting adventures. The novel became the dominant and most popular form of narrative art in the eighteenth and nineteenth centuries, as more and more writers devoted their lives to this art form, depicting and often satirizing contemporary life and morals. Throughout the nineteenth, and for most of the twentieth, century the novel became a powerful medium for modeling human nature and human society.

Narrative forms continue to be used to this day for modeling human behavior. Movies, TV programs, and the like are modern-day derivatives of the novel form. These attempt to model human behavior symbolically. Human motives are revealed obliquely by means of dialogue and through the minute depiction of character.

4.3.2 Tertiary nonverbal composite modeling

The Pythagorean Theorem discussed above (§4.3) is an example of a tertiary nonverbal composite form. All mathematical equations and theories are tertiary nonverbal composite models. Consider, as another example, the trigonometric ratios. In a right-angled triangle, the value of the *sine* of an acute angle, for instance, is equal to the length of the side of the triangle opposite the angle divided by the length of the hypotenuse. The sine varies in numerical value from 0 to 1 as the angle increases from 0° to 90°. Now, this ratio was known at first, not by theoretical proof, but by observing it as a characteristic of triangular figures. It was, in fact, the ancient Babylonians who established the measurement of angles in triangles, noting specific patterns. In the second century BC the Greek astronomer Hipparchus compiled a trigonometric table for solving triangles. At perhaps the same time, Indian astronomers had developed a trigonometric system based on the sine function. The earliest applications of trigonometry were in the fields of navigation, surveying, and astronomy, in which the main problem generally was to determine an inaccessible distance, such as the distance between the earth and the moon, or of a distance that could not be measured directly, such as the distance across a large lake.

Tertiary composite nonverbal modeling is not limited to the scientific domain; it can be found in all domains of human social life, from the routines that make up dance styles to the complex rituals that characterize courtship practices. Tertiary nonverbal modeling, however, is not unique to anthroposemiosis. It can be found in the zoosemiotic realm as well, albeit rarely. The best known example is the "wagging" dance of honeybees (chapter I, §1.4.2). In a classic paper published in 1923, the Austrian zoologist Karl von Frisch described how after a field bee discovers a new source of food, such as a field in bloom, it fills its honey sac with nectar, returns to the nest or hive, and performs a vigorous but highly standardized dance. As we saw previously (chapter II, §2.3.1), if the new source of food is nearby, within about 90 m of the nest or hive, the bee performs a circular dance, first moving about 2 cm or more, and then circling in the opposite direction. Numerous bees in the nest or hive closely follow the dancer, imitating its movements. Other bees then leave the nest or hive and fly in widening circles until they find the source. The imitative movements of the other bees constitute iconic modeling, but the circular dance in itself is obviously a tertiary model of the referent ([food source]). If the [food source] is farther away, the returning bee performs a more elaborate type of dance characterized, especially, by a series of intermittent movements across the diameter of the circle and by a vigorous, energetic "wagging" of its abdomen. Every singularized movement of this dance seems to have significance:

- The number of times the bee circles during a given interval informs the other bees how far to fly.
- Movement across the diameter in a straight run indicates the direction of the [food source]. If the straight run is upward, the source is directly toward the sun.
- If the straight run is downward, it signifies that the bees may reach the food by flying with their backs to the sun.
- If the straight run veers off at an angle to the vertical, it signifies that the bees must follow a course to the right or left of the sun at the same angle that the straight run deviates from the vertical.

The complexity of this dance language has paved the way for studies of tertiary modeling in other species. Some species are now known to have a variety of signals designed to facilitate social interactions. These appear, consequently, to have similar kinds of functions to the tertiary symbols that fuel and preserve human cultural life.

Take, for example, the symbolicity that characterizes the courtship displays of the bowerbird, of New Guinea and Australia—a passerine bird (of the family *Ptilonorhynchidae*). Remarkably, bowerbirds—so-named because the males build mating bowers, variously decorated, to attract females—put on displays that revolve around the use of inanimate objects. Males live apart from females for most of the year, but in breeding season gather together to compete for mates. Each male clears an area on the forest floor to which he attempts to attract females. He does this by placing shells, flowers, brightly colored berries, and even human-made objects. He might also build various kinds of structures: e.g., "maypoles" of sticks, often decorated with lichens and flowers, around a tree trunk; or tepee-like structures with a low entranceway in front of which is a "garden" of bright objects and flowers. Clearly, these "symbolic artifacts" are meant to appeal to the female, in much the same way as the giving of symbolic artifacts (flowers, jewelry, etc.) is intended to be an appeal strategy in human courtship. Similar courtship displays have been documented across a broad variety of bird species.

4.4. Tertiary cohesive modeling

Tertiary cohesive modeling involves symbolicity. The *intellective* representational activities discussed above of a mathematical nature are made possible by codes that provide the appropriate symbolic resources (numerical, geometrical, etc.) to represent mathematical relations in a cohesive fashion. Tertiary codes can also be discerned to underlie the symbolically-directed social behaviors that characterize daily interactions in a specific culture. *Social codes* (dress, gender, food, space, grooming, etc.), for instance, provide the symbolic resources for making messages about oneself and for regulating interpersonal activities. We will discuss intellective and social codes below.

4.4.1 Intellective codes

Intellective codes are those that have been designed to organize knowledge about some field, functioning as mental templates for understanding the world. A perfect example of an intellective code is that of trigonometry, which, as we saw above (§4.3.2), is based on the relations between the sides of triangles.

The six trigonometric functions are defined in terms of a given acute angle in a right triangle:

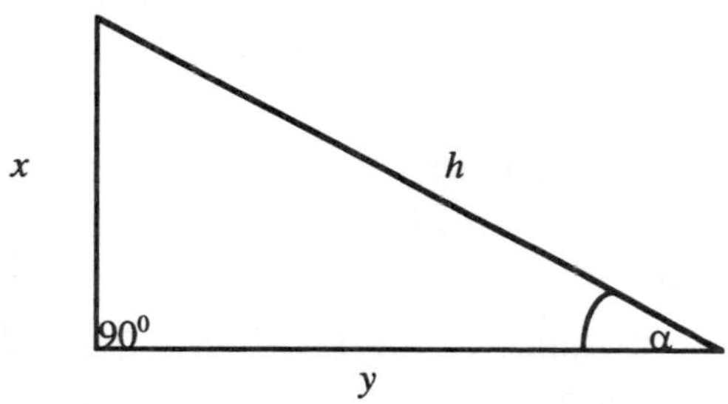

Figure 47. Right-angled triangle used for defining the trigonometric functions of an acute angle (α)

The *sine* (sin) is defined as the ratio of the opposite side to the hypotenuse, x/h; the *cosine* (cos) as the ratio of the adjacent side to the hypotenuse, y/h; the *tangent* (tan) as the ratio of the opposite side to the adjacent side *x/y*; the *cotangent* (cot) as the ratio of the adjacent to the opposite side, *y/x*, the *secant* (sec) as the ratio of the hypotenuse to the adjacent side, *h/y*, and the *cosecant* as the ratio of the hypotenuse to the opposite side, *h/x*. For any angle the numerical values of the trigonometric ratios can be easily approximated by drawing the angle, measuring, and then calculating the ratios.

While this code appears to have little relevance to real-world situations, the remarkable thing is that it can be applied to solve paractical problems in engineering, navigation, construction, etc. By representing an unmeasurable distance as one side of a triangle on some visual composite model of a situation, measuring other sides or angles of the triangle of the model, and applying the appropriate trigonometric formulas to the model, the distance can be easily determined.

Take, as a simple example, the following problem:

From a lighthouse, a keeper spots a boat at an angle of depression of 43^{0}. If the keeper's eyes are 39 meters above the water, how far is the boat from the base of the lighthouse.

As this problem shows, trigonometry makes solving what would otherwise be an intractable situation a simple affair:

- First, we draw a diagram that, in its bare outline, represents what is known about the scene. This is, of course, a visual composite model of the scene.
- Assuming that the lighthouse is standing upright, we can draw it as a straight line, 39 meters in length, above the water (= one side of the triangle).
- Since we do not know the distance of the boat from the base of the lighthouse, we can represent it with the letter *d* (= another side of the triangle).
- By joining the point where the boat is seen at distance *d* to the keeper's eyes on top of the lighthouse (= the third side of the triangle), we can thus complete the triangular model of the scene.
- This model is shown in figure 48.

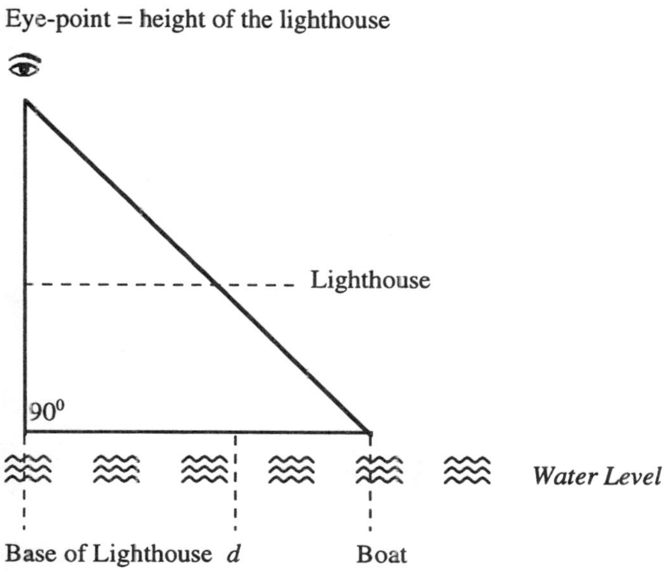

Figure 48. Diagrammatic model of the distance of a boat from a lighthouse

Now, since the angle of depression, i.e., the angle that is measured down from the horizontal to the boat, is given as 43^0, we can mark it on the diagram in its appropriate spot (see figure 49). Because the horizontal is parallel to the water, i.e., to line *d*, the angle between *d* and the hypotenuse is, by a proposition of Euclidean geometry, also 43^0. We can now consider the triangle on its own, without reference to the real-world problem at hand, because, like any symbolic representation, we can examine its form on its own terms and perform the requisite operations on it. The *tan* of the angle 43^0 is the ratio $39/d$ (the ration of the opposite side to the adjacent side). We can now set up an equation as follows, and solve for *d*, since we know (from the appropriate trigonometric function table) that tan 43^0 is 0.932515:

$$\begin{aligned} \tan 43^0 &= 0.932515 \\ 39/d &= \tan 43^0 \\ 39/d &= 0.932515 \\ d &= 39/0.932515 \\ d &= 41.8 \end{aligned}$$

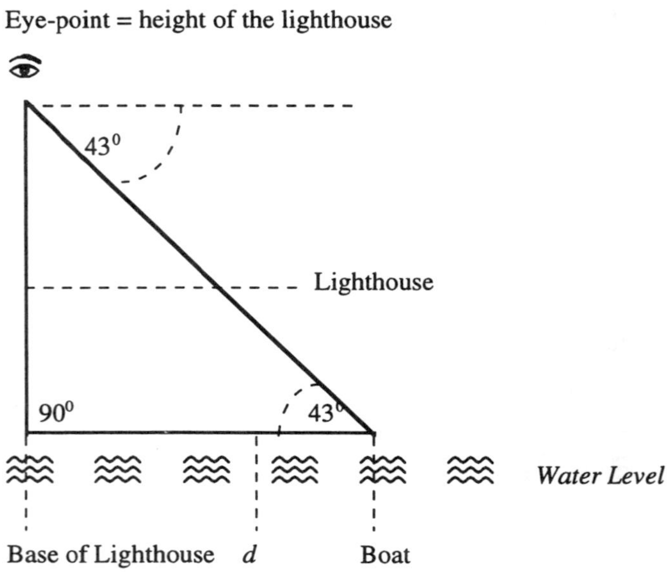

Figure 49. Diagrammatic model of the distance of a boat from a lighthouse showing angle of depression

With this simple procedure we have measured the distance of the boat from the base of the lighthouse as (approximately) 41.8 meters *intellectively*, i.e., without having to do it physically.

4.4.2 Social codes

Social codes are interconnected with each other creating an overarching interpersonal regulatory system based on symbolicity. In this section we will discuss briefly two common types of social codes: clothing and food.

At a biological level, *clothes* have a very important function indeed–they enhance human survivability considerably. They are, at this denotative level, human-made "additions" to protective bodily hair and skin thickness. This is why clothing styles vary in relation to different climatic zones. But in social settings the separate items of clothing cohere symbolically into the various *dress* signifiers (from Old French *dresser* 'to arrange, set up') that inform people how to present themselves in society. In terms of the dimensionality principle (chapter I, §1.2.1), clothes denote, first, bodily protection (i.e., they extend bodily protective functions); second, they take on specific connotative meanings in social settings; and, third, these meanings form the basis of the symbolism that is associated with the clothes worn during certain ceremonies and rituals.

Dress can even be used to lie about oneself: e.g., con artists and criminals can dress in three-piece suits to look trustworthy; a crook can dress like a police officer to gain a victim's confidence, and so on. To discourage people from deceiving others through clothing, some societies have even enacted laws that prohibit fake dressing. In ancient Rome, for instance, only aristocrats were allowed to wear purple-colored clothes; in medieval Europe peasants were required to wear their hair short because long hair was the privilege of the aristocracy; in many religiously-oriented cultures differentiated dress codes for males and females are regularly enforced; and the list could go on and on.

This phenomenon is found in other species as well, where shedding is, in effect, the zoosemiotic counterpart to changing clothes. The mayfly, an insect that often emerges in great numbers from lakes, streams, and rivers, usually spends one to three years as an underwater creature, breathing by means of gills and feeding on microscopic plant life. After a while mayflies emerge from their nymphal skins on the water surface and fly to nearby plants, where they shed their downy, waterproof skins. This changes their life role. Now fully adult, they cannot feed, but instead they form male and female swarms that mate over water. After mating, the males die; the females live a few more hours, depositing the eggs in water, giving birth to the next generation of nymphs.

The coloration of plants is, in effect, vegetative clothing. In flowers this takes the form of different pigments: fat-soluble pigments contained in chromoplasts and water-soluble pigments contained within the vacuoles of the epidermal cells of the petals. Scattered throughout the epidermis of plant leaves are pairs of bean-shaped cells, called guard cells, which contain chloroplasts, tiny granules filled with the green pigment chlorophyll. These enable leaves to carry on photosynthesis because they allow them to absorb carbon dioxide and sunlight. Chlorophyll gives leaves their green color. The presence of additional pigments causes other colors such as red and purple. In temperate regions of the world, the leaves of some plants change color in autumn, when chlorophyll production decreases and the other pigments become visible.

In the human world, clothing generates connotata ([sexuality], [gender], [class], [role], etc.) across the world, as well as religious, military, and other kinds of social symbolism (i.e., symbolic meanings). This is why clothing is an intrinsic component of various rites, rituals, and traditions that give a sense of social cohesion to people across the world. The various items of clothing that priests, rabbis, shamans, and other clerics wear at certain times of the year, for instance, are all symbolic of common ethical and moral values.

Like clothing, *food* is much more than just nourishment and sustenance in human affairs. We eat, first and foremost, to survive. Denotatively, food is survival substance. But, at the level of culture, food items are more symbol than substance. The anthropologist Claude Lévi-Strauss (1964) traced the origin of food symbolism to the emergence of cooking technology. Cooked food is aliment that has been transformed into something more than a survival substance. According to Lévi-Strauss this transformation was accomplished by two processes–roasting and boiling–both of which were among the first significant technological advances made by humans. Roasting is more primitive than boiling because it implies a direct contact between food and fire. But boiling reveals an advanced form of technological thinking, since the cooking process in this case is mediated by a pot and a cooking process. This dichotomy entails an array of symbolic distinctions. In some parts of the world it has been enshrined into the social system to symbolize interpersonal and class relations. In the Hindu caste system, for instance, the higher castes may receive only raw food from the lower castes; whereas the lower castes are allowed to accept any kind of cooked food from any caste.

Early food symbolism still reverberates in modern-day social practices and traditions. The world's religious ceremonies, for instance, are centered on food symbolism. The *raison d'être* of the Catholic Mass, for instance, is to partake symbolically of the consecrated body and blood of Christ; specific types of food are served and eaten traditionally at Thanksgiving, Easter, Christmas; feasts such as weddings Bar Mitzvahs, etc. are centered on food symbolism; and the list could go on and on. This is why we talk of the *bread of life*, of *earning your bread*, of *sacrificial lambs*, and the like.

Food codes also determine edibility. Apart from those which have a demonstrably harmful effect on the human organism, the species of flora and fauna that are considered to be edible or inedible is very much an arbitrary cultural decision. The perception of a food's edibility has a basis in symbolism, not digestive processes. We cannot get nourishment from eating tree bark, grass, or straw. But we certainly could get it from eating frogs, ants, earthworms, silkworms, lizards, and snails. Most people in Anglo-American culture would, of course, respond with disgust and revulsion at the thought of eating such potential food items. However, there are cultures where they are not only eaten for nourishment, but also considered to be delicacies. In our culture rabbits, cats, and dogs are classified as "household pets", and this predisposes us to perceive cooked rabbit, cat, and dog meat as "inedible". However, we routinely eat bovine meat (beef steaks, hamburgers, etc.), lamb meat, and poultry meat, with few negative perceptions. In India, on the other hand, a cow is classified as "sacred" and, therefore, as "inedible"–incidentally, this is the basis of the expression *sacred cow*. Anglo-American culture does not classify foxes or dogs as edible food items; but the former is reckoned a delicacy in Russia, and the latter a delicacy in China.

Food symbolism also underlies how people prepare food and when and how they eat it. Many Christians say grace before starting a meal together; Jews say special prayers before partaking of wine and bread. At a formal meal, the order in which dishes are presented, what combinations can be served in tandem, how the foods are to be placed on the table, who has preference in being served, who must show deference, who does the speaking and who the listening, who sits where, and what topics of conversation are appropriate are all steeped in cultural symbolic history and tradition. Eating events are so crucial to the establishment and maintenance of social relations and harmony that there exists virtually no culture that does not assign an area of the domestic abode to eating functions and ceremonies. All cultures, moreover, have a discrete set of table rituals and manners that are inculcated into the members of the culture from birth. If one does not know the *table-manner code* of a certain culture, then one will have to learn it in order to continue living in that culture without censure and disapprobation.

4.5 Tertiary connective modeling

Tertiary connective modeling involves linkages from which a symbolic form can be extracted. We have designated this type of symbolic form a *meta-symbol* (chapter I, §1.6.2). In this section we will look briefly at the formation of meta-symbols and at the metaphorical basis of common discourse.

4.5.1 Meta-symbols

A *rose* is used as a symbol for *love* in Western culture because the distinctive features associated with a *rose*–[sweet smell], [red color], and [plant]–constitute source domains for conceptualizing *love*: [love = sweet smell] ("Her kisses are *sweet*"), [love = red color] ("Her *red* lips are inviting"), [love = plant] ("Our love has been *blossoming* for a number of years"). Now, it is easy to see how the [rose = love] meta-symbol came about:

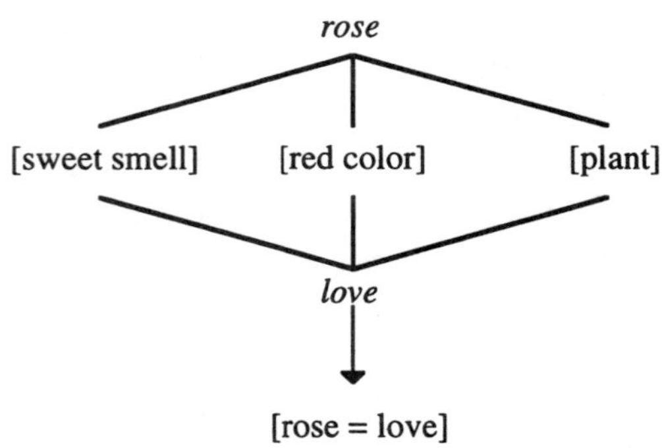

Figure 50. Derivation of the meta-symbol [rose = love]

This is why roses are given to loved ones on romantic occasions, why they are used as images of love in narratives, and so on. The [rose = love] linkage is a meta-symbol derived from the fact that

[love] has been associated with source domains that are descriptive of a physical rose ([sweet smell], [red color], [plant]).

Take, as another example, the notion of *personality*. We start by considering the following metonymic metaforms which deliver the concept of *person*:

[body part = the person]

132. Get your *butt* over here!
133. He's the *head* of that organization.
134. We don't hire just *muscles*.

[the face = the person]

135. He's just another handsome *face*
136. Put on a happy *face*!
137. We must talk, *face* to *face*.

[clothing apparel = the person]

138. What *hat* are you wearing today?
139. We don't hire *blue jeans* around here.
140. We need more stiff *collars* in this office.

[body style = the person]

141. He's just another *shaved head*.
142. Ignore her, she's *slow of foot*.
143. Ignore him, he's a *swaggerer*.

The constant juxtaposition of such metaforms in common discourse and other kinds of representational activities produces, cumulatively, a *meta-symbol* of personality, which has the form [physical appearance = personality]. This model came about, clearly, because the things that define physical appearance–[body parts],

[face], [clothing apparel], [body style]–also constitute source domains for [personality]:

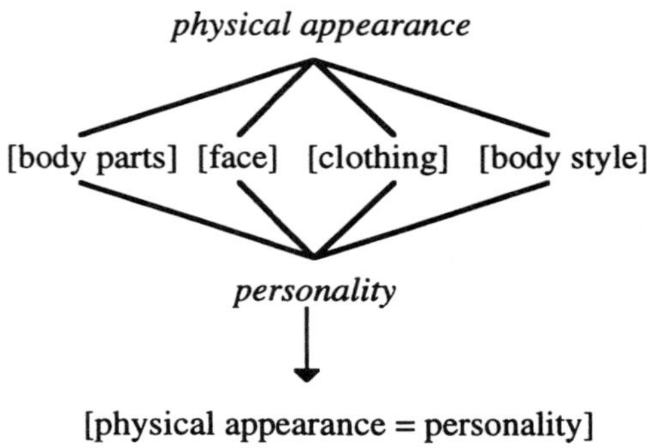

Figure 51. Derivation of the meta-symbol [physical appearance = personality]

This is the reason why, for instance, portraits are made of the face, why body image is crucial in presenting an appropriate *persona* to the social milieu, why clothing styles are associated with personality and lifestyle, and so on.

Meta-symbols are found throughout the signifying order, and are traces to a culture's past. A common expression like "He has fallen from grace" would have been recognized instantly in a previous era as referring to the Adam and Eve story in the Bible. Today we continue to use it with only a dim awareness (if any) of its Biblical origins. Expressions that portray [life] as a [journey]–"I'm still a *long way* from my goal", "There is no *end in sight*", etc.–are similarly rooted in Biblical meta-symbolism. As the Canadian literary critic Northrop Frye (1981) aptly pointed out, one cannot penetrate such expressions, and indeed most of Western literature or art, without having been exposed, directly or indirectly, to the original Biblical meta-symbols.

The meta-symbolic link to the past is also evident in proverbial language:

144. You've got too many fires burning (= *advice to not do so many things at once*)

145. Rome wasn't built in a day (= *advice to have patience*)
146. Don't count your chickens before they're hatched (= *advice to be cautious*)
147. An eye for an eye and a tooth for a tooth (= *equal treatment is required in love and war*)

These result from the fact that the distinctive features of concrete concepts, which are implicit ones in proverbial language, are also the source domains for certain abstract notions. Take, for example, the "Rome wasn't built in a day" saying. The implicit concrete feature associated with building great cities (such as Rome) is: [takes a long time]. Now, this is perceived as being true of anything that has lasting value ([learning], [loving], etc.):

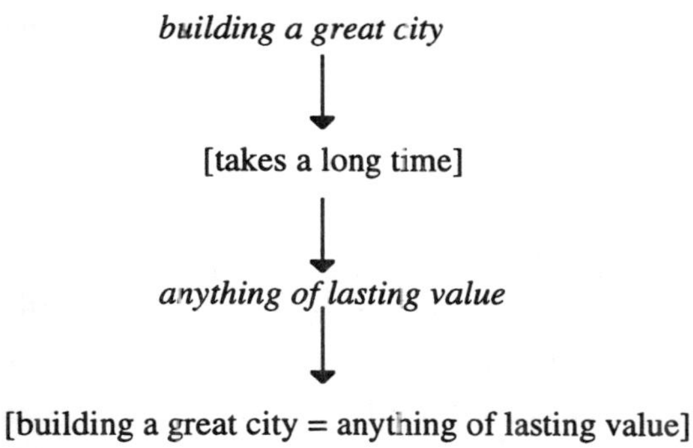

Figure 52. Derivation of the meta-symbol [building a great city = anything of lasting value]

Every culture has similar proverbs, aphorisms, and sayings. They constitute a remarkable code of ethics and of practical knowledge that anthropologists call "folk wisdom". Indeed, the very concept of *wisdom* implies the ability to apply proverbial language insightfully to a situation.

The use of meta-symbols extends to scientific reasoning. Science often involves things that cannot be seen–atoms, waves, gravitational forces, magnetic fields, etc. So, scientists use their metaphorical know-how to get a look, so to speak, at this hidden matter. That is why waves are said to *undulate* through empty space like water waves ripple through a still pond; atoms to *leap* from one quantum state to another; electrons to *travel in circles* around an atomic nucleus; and so on. The poet and the scientist alike use meta-symbols to extrapolate a suspected inner connection among things. These are slices of truth; they are evidence of the human ability to see the universe as a coherent organism. When a meta-symbol is accepted as fact, it enters human life, taking on an independent conceptual existence in the real world, and thus can suggest ways in which to bring about changes in and to the world. Euclidean geometry, for instance, gave the world a certain kind of visual meta-symbolic structure for millennia–a world of relations among points, lines, circles, etc. But this structure was changed to suit new conditions and ideas when Nicholas Lobachevski (1793-1856) literally imagined that Euclid's parallel lines would "meet" in some context, such as at the poles of a globe. As physicist Robert Jones (1982: 4) aptly puts it, for the scientist metaphor serves as "an evocation of the inner connection among things".

4.5.2 Discourse

Discourse is a pliable tertiary modeling system. It is interesting to note, however, that the "meaning flow" that characterizes most discourse situations is metaphorical. The evidence for this has become quite substantive. For instance, one recent study (Danesi 1999b) has shown that meaning flow in discourse is shaped largely by a syntagmatic chain of metaforms (i.e., by the elements of their source domains and sub-domains)–a finding which suggests that discourse unfolds primarily through a "circuitry" of source domains through which interlocutors "navigate mentally", so to speak.

The following brief stretch of recorded conversation between two students on the University of Toronto campus (Danesi 1999b) shows how the [human personality = perceived physical features of animals] metaform (chapter II, §2.5) shaped the pathways of one of the circuits of their conversation:

Student 1: You know, that prof is a real *snake*.
Student 2: Yeah, I know, he's a real *slippery* guy.

Student 1: He somehow always knows how to *slide* around a tough situation.

Student 2: Yeah, tell me about it! Keep away from his courses; he *bites*!

The circuit that this metaform triggered in that conversation is represented in figure 53 below:

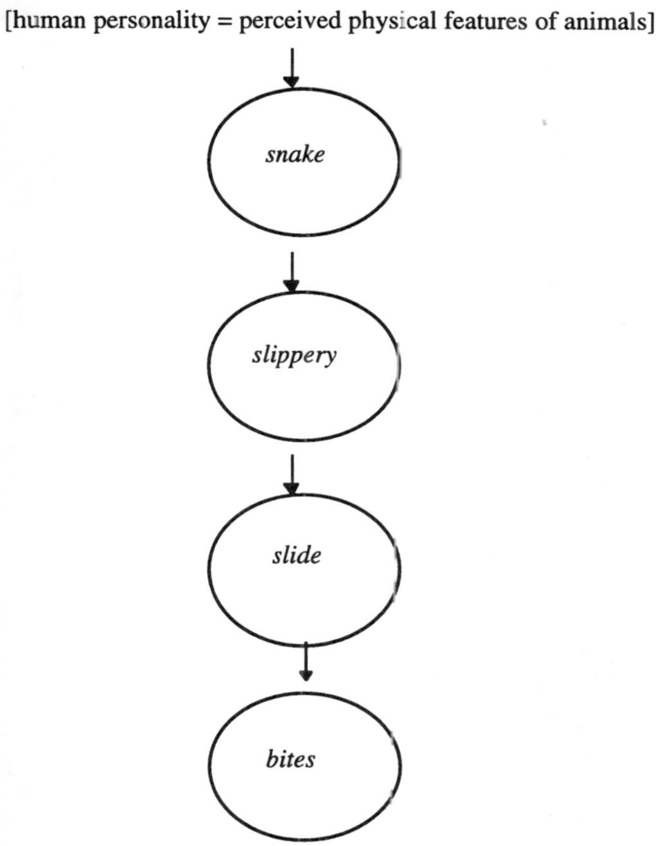

Figure 53. Discourse circuit triggered by [human personality = perceived physical features of animals] metaform

Sometimes, the circuit is shaped by a series of metaforms, which are interconnected to each other in the discourse pathway. In one conversation about *ideas*, an interlocutor made use of the following sequence of source domains (Danesi 1999b): [ideas = seeing] → [ideas = food] → [ideas = fashion] → [ideas = persons]:

> *I do not see how anyone can swallow his ideas, especially since most of them have gone out of fashion, and thus are dying.*

The circuit in this case is shown in figure 54 below:

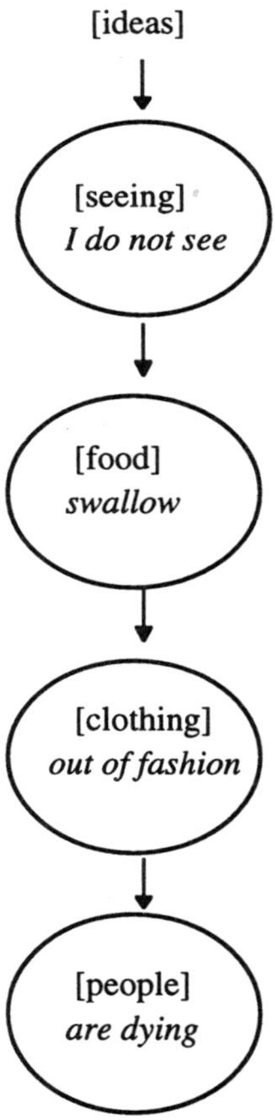

Figure 54. Discourse circuit triggered by various source domains associated with [ideas]

Like an organism, discourse is a highly adaptive and context-sensitive instrument, susceptible to the subtle connotative nuances that the situation calls for. In traditional theories of discourse, denotation is considered to be the primary shaper of the cognitive flow of meaning, and connotation only a secondary, context-dependent option within this flow. But this type of "dictionary" model of meaning yields very little insight into the true nature of verbal communication. In actual discourse, it is the connotative (metaphorical) dimension of connective modeling that guides the "navigation" through the discourse circuit. This is why someone who studies a foreign language has, initially, little or no access to such circuits, given that language teaching tends to be based on denotative models of meaning. The foreign language learner can rarely be a participant in real discourse situations until s/he has acquired the underlying metaformal structures that shape discourse circuitry. Michel Foucault (1976) characterized such circuitry as consisting of an endless "interrelated fabric" in which the boundaries of meanings are never clear-cut. Every signifier is caught up in a system of references to other signifiers, within a network of distributed signifieds.

Chapter V
Systems analysis

The scientific mind does not so much provide the right answers as ask the right questions.

Claude Lévi-Strauss (1908-)

5. Introductory remarks

The purpose of this book has been to show how the many *forms of meaning* produced by singularized, composite, cohesive, connective modeling are the products of three distinct yet interconnected and overlapping modeling systems. Some of these forms are unique to human modeling, others cut across the semiotic realms (phyto-, zoo-, and anthroposemiotic). The essence of *Systems Analysis* (SA), as we have argued primarily through illustration, is that the specific kinds of modeling phenomena that are characteristic of the various species give the analyst access to the workings of their particular modeling system(s). The purpose of SA is to provide a framework for investigating *how* models are constructed, *what* their species-specific functions are, *why* they came about, and *how* they generate forms of meaning.

Recall from the opening chapter (chapter I, §1.2.3), that the point-of-departure for conducting this line of inquiry originated in the study of symptoms–the natural forms produced by the body. In *The Science of Medicine*, Hippocrates stated: "What escapes our vision we must grasp by mental sight, and the physician, being unable to see the nature of the disease nor to be told of it, must have recourse to reasoning from the symptoms with which he is presented", and thus that the essence of diagnosis is to "deduce of what disease they (the symptoms) are the result, what has happened in the past and to prognosticate the future course of the malady" (cited in Chadwick and Mann 1950: 87-89). Shortly thereafter Galen (130?-200? AD), attempted to provide prognostics, wherever feasible, with a scientific underpinning, i.e., to base forecasts on actual observations. He was able to do this because he practiced dissection and experiment, whereas Hippocrates studied disease as a naturalist. Galen "dared to modify nature as a scientist" (Majno 1975: 396), and can therefore be regarded as a subtle founder of clinical semiotics (Neuburger 1906: 385). But he can also, very likely, be reckoned the first true semiotician, in the modern sense of the word.

As discussed in the opening chapter (§1.1), a symptom is a natural primary model, produced by a particular bodily process. But its interpretation in medical terms is indexical, since its form is perceived to be a signifier pointing to some area or aspect of physiology or bodily morphology. Symptoms are, in effect, indexical promptings of the body crying out for an explanation (Polunin 1977: 91) and, moreover, in the human realm they are assigned a subjective, culturally-based interpretation that influences how they are experienced. Consider pain. As the ancient physicians also knew, pain thresholds are influenced by non-physical factors, such as the unique past history of the individual and the culture's appraisal of a certain physical condition. As the psychologist Melzack (1972: 223) has observed, in North American culture "childbirth is widely regarded as a painful experience"; but there are "cultures in which the women show virtually no distress during childbirth". This does not mean that North American women are making up their pain. It means that childbirth is perceived as possibly endangering the life of the mother, and young females learn to fear it in the course of growing up.

As Friedrich J. K. Henle, the illustrious nineteenth century German anatomist and physiologist pointed out, subjective reactions to pain have no physical cause for existence, nor is there any organic basis for the experience of pain. It results from "impressions stored up in the memory centers, which are recalled by the proper associations aroused" (Behan 1926: 74), which is to say that pain is a consequence of personal and social factors, not organic ones. Certain symptoms are private experiences, housed in no identifiable site, but in an isolated annex that humans usually call "the Self". Symptoms such as these tend to be expressed by vocal means, such as groans, tones, or various expressions, which may or may not be coupled with gestures, ranging in intensity from frowns to writhings. An exceedingly knotty problem, which can barely be alluded to here, arises from the several meanings of Self and how these relate to the matter of symptomatology. The biological definition hinges on the fact that the immune system does not respond overtly to its own Self-antigens; there are specific markers that modulate the system generating antigen-specific and idiotype-specific cell lines (Sebeok 1979: 263-267).

The craft of interpreting symptoms has a significance far exceeding the physician's day-to-day management of sickness. As Hippocrates had already anticipated, its success derives from its psychological power, which critically depends on the practitioner's ability to impress h/er skills on both the patient and their joint environment (the audience gathered in h/er workshop, which may consist of the patient's family and friends, as well as the physician's colleagues and staff). According to recent medical thinking, the contemporary preoccupation with diagnosis–i.e., the doctor's perceived task of explaining the meaning of the patient's condition–rests in the final analysis with the doctor's self-assigned role as an authenticated expositor and explicator of the values of contemporary society. Disease is thus elevated to the status of a moral category.

In a fundamental sense, the study of all modeling phenomena is a study of the symptomatology between body, mind, and culture–just like the ancient physicians maintained. SA is grounded in this tradition. In chapters II, III, and IV we laid the groundwork for SA by discussing the three types of modeling systems, selectively documenting the multifarious manifestations of semiosis in and across various species, so as to establish a taxonomy of notions, principles, and procedures for understanding how modeling unfolds. In this chapter, we will tie some loose theoretical and methodological strings together.

5.1 The framework for systems analysis

The main tasks of SA are: (1) to determine what constitutes a model in animal behavior, (2) to what modeling system it pertains (PMS, SMS, TMS), (3) what kind of modeling activity it manifests, and (4) what its function is. These tasks are guided by several key notions. First, there is the notion of dimensionality which posits three distinct but interconnected types of models: (1) a *primary* model, which is a simulacrum of a referent; (2) a *secondary* model, which is either an extension of a simulacrum or an indexical form; and (3) a *tertiary* model, which is a symbolically-devised form of some kind. Second, there is the notion of stability vs. pliability which claims that a model (natural or artificial) can be *stable* (as in a written text) or *pliable* (as in oral conversation): stable models are fixed and relatively permanent or invariable; pliable ones are temporary and adaptive to the dynamics of a situation. Third, there is the notion which posits that the *form* a model assumes can be *singularized*, *composite*, *cohesive*, or *connective*, providing clues as to the nature of the referent or referential domain that it encodes. Fourth, there is the notion of interconnectedness, whereby the modeling system deployed will vary according to the nature of the referent, the function of the model, and the situation in which the modeling act occurs. Fifth, a distinction is made among semiosis, modeling, and representation: *semiosis* is the neurobiological capacity to produce forms (signs, texts, etc.), *modeling* is the channeling of the *semiosic* capacity towards a *representation* of some referent (the actual act of creating a form). Sixth, there is the notion that all models possess the same structural features (paradigmaticity, syntagmaticity, etc.). Finally, there is the notion that modeling reveals how the brain carries out its work of transforming sensory forms of knowing into internal forms of thinking and external forms of representation: a specific external *model* is thus considered to be a "cognitive trace" to the form a concept assumes in the mind, and since concepts depend on how they are modeled it has been argued throughout this book that the *form* that *knowledge* takes depends on the type of *modeling* used.

5.1.1 Biosemiotics vs. sociobiology

At this point, it is essential to differentiate between the biosemiotic approach to the study of modeling and communication across species and the one called *sociobiology* which, although it shares many features with SA, is grounded on two flaws–the inability to distinguish between the genetic and the semiosic, and the inability to see an interconnection between internal models and external ones.

Sociobiology combines information and techniques from the social and biological sciences in order to study and explain the evolutionary and cultural bases of animal behavior. The sociobiological account of cultural genesis starts with a consideration of the origin of life, which occurred when a single-celled organism emerged with the capacity to reproduce itself. The next stage consisted in the development of a more complex cell–the basis of all higher life forms including human body tissues. The step after that occurred when larger multicelluar organisms (flatworms, crustaceans, etc.) with the capacity to develop more complex organs like eyes and brains came into existence. The last giant step consisted in the emergence of the reflective human brain. Sociobiologists attempt to describe what caused the leap from largely genetically programmed behavior to reflective thought in terms of a gene-culture "coevolution" process (e.g., Lumsden and Wilson 1983). This process was purportedly triggered in *Homo Habilis* hominids after they had learned how to use their hands to make tools. They were small creatures with a human body and a brain similar to that of an intelligent ape. They lived in groups as hunter-gatherers on the Savanna plains of Africa. Threatened by larger mammals, but desperately needing to catch game in order to survive, they had to learn how to act cooperatively, to think logically, and to communicate among themselves in some coherent and functional manner. This situation led, it is claimed, to the institution of communal rules for hunting, food sharing, division of labor, mating, etc. Theirs was the earliest culture of humanity. The subsequent *Homo Erectus* and *Homo Sapiens* stages became even more "coevolutionary", because culture apparently took over more and more of the load in the evolutionary thrust forward of humanity. *Homo Erectus* lived from about 1.8 million to 30,000 years ago. This hominid had a larger brain, measuring up to 1150 cc, and a rounder cranium than earlier hominids, enabling it to do many things its ancestors had never been able to do: to make tools, to hunt systematically, to construct campsites, to use fire, etc. The most characteristic of the tools made by *Homo Erectus* was a teardrop-shaped hand ax.

Homo Erectus species, so the story goes, migrated from Africa to parts of Asia and Europe, where it gradually evolved into *Homo Sapiens*, within the past 200,000 years through interbreeding and cultural admixture. Between 200,000 and 300,000 years ago, *Homo Sapiens* appeared. Finally, *Homo Sapiens Sapiens* appeared around 100,000 years ago. There is some disagreement among scientists on whether the hominid fossil record shows a continuous evolutionary development from the first appearance of *Homo Sapiens* to modern humans. Whatever the outcome of the disagreement, the evidence shows that early *Homo Sapiens* groups were highly efficient at exploiting the sometimes harsh climates of Ice Age Europe, that they buried their dead deliberately, that they communicated with vocal speech, and that they had a highly-evolved tool-making technology–all of which were indicative that culture was becoming a primary force in the evolution of modern humans.

In the sociobiological account, the human mind was generated by a symbiotic partnership between genes and culture. As cultures became more complex, so did the human mind, which gradually allowed humans to make choices that conferred upon them greater survival and reproductive abilities. Gene evolution gradually gave way to cultural evolution. The body's survival mechanisms were eventually replaced by those of the mind.

The problems that sociobiologists seek to resolve can be traced ultimately to the problems that Darwin attempted to resolve. These were articulated in a synthetic fashion by the American biologist E. O. Wilson in 1975, in a work that has become the cornerstone of sociobiology. At about the same time, British biologist Richard Dawkins (1976) came forward to seemingly provide the "missing link" to account for the change from genetic to cultural evolution, by proposing the concept of *meme*–a word he coined in direct imitation of the word *gene*. Dawkins defined *memes* as replicating patterns of information (ideas, laws, clothing fashions, art works, etc.) and of behavior (marriage rites, love rituals, religious ceremonies, etc.) that people inherit directly from their cultural environments. Like genes, memes involve no intentionality on the part of the receiving human organism. Being part of culture, the human being takes them in unreflectively from birth, and then passes them on just as unreflectively to subsequent generations, allowing them to improve adaptively over preceding generations. Memes, Dawkins claims, are responsible for cultural progress, advancement, and betterment, having become *the* primary agents in the human species' evolutionary thrust forward.

Dawkins' case is, at its core, a deceptive one. *Genes* can be identified and separated from organisms, and then studied, altered, and even cloned physically. That is a scientific fact. A *meme,* on the other hand, is no more than Dawkins' own term for what we have called a *model* in this work. There is no empirical way to verify the reality of memes, as defined by Dawkins; they can only be talked about as if they existed. But it is possible to study the structure of mind in the structure of models, as we have done throughout this book. It is impossible to go here into the arguments that have been put forward for and against the sociobiological paradigm. Suffice it to say that sociobiology has failed to distinguish between genetic and semiosic processes, and it has failed to see the interconnection between the concepts that people have and the models they use to encode them.

Recall Popper's theory of culture (chapter IV, §4.1.2). World 1 in Popper's scheme corresponds to the pre-conscious world of instincts and sense impressions. It inheres in what can be called an "immunologic" or "biochemical" Self with semiotic overtones; whereas World 2 corresponds to the "semiotic" Self (Sebeok 1979: 263-267); and World 3 to the representational end-products of anthroposemiosis–symbols, institutions, cultural practices, etc. The genetic code of World 1 is qualitatively distinguishable from the semiosic codes that govern Worlds 2 and 3 states of consciousness, although there is an interconnectedness among the worlds, as we have seen, through extensional modeling. In effect, the biological, semiosic, and representational dimensions of Self-hood are intertwined in a "modeling symbiosis"; in no way does one dimension (e.g., the cultural) ever replace the other (e.g., the genetic).

As mentioned, the key figure behind sociobiological theory and research is the American biologist E. O. Wilson (1929-), known for his work tracing the effects of natural selection on biological communities, especially on populations of insects, and for extending the idea of natural selection to human cultures (e.g., Wilson 1975, 1979, 1984). Wilson claims that the psychological capacities and social behaviors that humans manifest are genetically based and that evolutionary processes favor those that enhance reproductive success and survival. Thus, characteristics such as heroism, altruism, aggressiveness, and male dominance, for instance, should be understood as evolutionary outcomes, not in terms of historical, social, or psychic processes. Moreover, he sees the creative capacities undergirding language, art, scientific thinking, etc. as originating in the same pool of genetic responses that help the human organism solve physical problems of survival and species continuity. As he has stated rather bluntly, "no matter how far culture may take us, the genes have culture on a leash" (in Wilson and Harris 1981: 464).

If there is any substance to Wilson's claim that culture is a reflex of the same genetic responses that have helped the human species adapt physically, then one can legitimately ask: What do such things as paintings, music compositions, marriage rites, burial rites have to do with survival or reproductive success? Animals, as we have seen in previous chapters, are endowed by Nature with the capacity to use and respond to signals for survival. Human beings also emit many types of signals in response to certain stimuli; but the real power of anthroposemiosis lies in its ability to transcend stimulus-based signaling behavior. This in no way implies that animals do not communicate effectively, nor that they may lack sophisticated modeling systems (e.g., Griffin 1981). We have provided various striking examples of these in previous chapters. But there is no evidence to suggest that animals are capable of "semiotic Self" and "social Self" experiences in the way that humans are. The available ethological evidence suggests that these are exclusively human. The lesson in all this is saliently obvious: as the ethologists discovered at mid century, each organism must be approached in terms of its different anatomical, sensory, and intellective structures.

The idea that there is a biological basis to conscious social behaviors is, of course, partially true; but it is not totally true. The claim that sapience resulted from natural selection is not only oblivious of human history, but also without any empirical foundation. The instinctive need to know seems to have a single goal–allowing people to solve similar basic physical and moral problems the world over. The signs, texts, codes, and metaforms that humans create, no matter how strange they might at first seem, have universal structural properties that allow people everywhere to solve similar life problems.

5.1.2 Carrying out systems analysis

The foregoing discussion was meant to emphasize that, although the two may seem to have similar goals, biosemiotics and sociobiology are actually quite dissimilar in terms of what they aim to investigate and how they go about doing so. As discussed and illustrated throughout this book, SA is guided by several key principles and notions. These are recalled summarily here for the sake of convenience.

The *modeling principle* asserts that representation is a de facto modeling process. As a corollary, the *representational principle* posits that the kinds of knowledge of which human beings are capable are indistinguishable from how they are represented. The form that a scientific theory assumes, for instance, will determine what is known about a certain phenomenon. Take, for example, the theory of parallels in geometry. One of the postulates of Euclid's (c. 300 BC) geometry states that through a point outside a given line it is possible to draw only one line parallel to the given line, i.e., one that will never meet the given line no matter how far the lines are extended in either direction. For many centuries mathematicians believed that this postulate could be proved on the basis of the remaining postulates; but all efforts to discover such a proof were fruitless. Nevertheless, it made "visible sense", because Euclidean geometry is a *representational system* that *models* geometric phenomena on the plane (e.g., a surface); and on a open plane, it can indeed be "seen" that parallel lines do not meet, no matter how far they are extended.

In the first part of the nineteenth century the German mathematician Carl Friedrich Gauss (1777-1855), the Russian mathematician Nikolai Ivanovich Lobachevski (1793-1856), and the Hungarian mathematician János Bolyai (1802-1860) independently demonstrated the possibility of coming up with a different representational system in which Euclid's postulate of the unique parallel was replaced by a postulate stating that through any point not on a given straight line an infinite number of parallels to the given line could be drawn. Later, around 1860, the German mathematician Georg Friedrich Bernhard Riemann showed that a geometry in which no parallel lines occurred was equally possible. The structural details of these two types of non-Euclidean systems are complex, but both can be demonstrated by means of simple models. The Bolyai-Lobachevski system, often called hyperbolic non-Euclidean geometry, describes the geometry of a plane consisting only of the points on the inside of a circle in which all possible straight lines are chords of the circle. In this plane, an infinite number of parallels can be drawn through a point in the circle, because they are "contained" by the circumference of the circle, and thus cannot meet. Riemannian, or elliptic non-Euclidean geometry, is the geometry of the surface of a sphere in which all straight lines are great circles. On a globe, for instance it is impossible to draw any pair of parallel lines. For comparatively small distances, Euclidean geometry and the non-Euclidean geometries are essentially equivalent. However, in dealing with astronomical space and such problems of modern physics as relativity and the theory of wave propagation, non-Euclidean geometries give a more precise description of the observed phenomena than does Euclidean geometry.

This chapter in the history of geometry makes it saliently obvious that modeling and knowing are intrinsically intertwined in human life. Euclidean geometry constitutes a "body of knowledge", as the expression goes, about certain types of phenomena that occur in a specific physical world; non-Euclidean geometries constitute different bodies of knowledge about certain types of phenomena that occur in different kinds of physical worlds. In a phrase, geometric knowledge is indistinguishable from how it is represented.

In SA, the species-specific forms of knowing are seen as manifest in the modeling behaviors of the species. Access to how a species knows something, therefore, is through the modeling system it possesses. Primary modeling, for instance, is "knowing through simulation". It is anchored in osmosis and mimesis. Secondary modeling, on the other hand, is "knowing through extension and indication". This implies that the SMS does its handiwork, by and large, after the PMS has completed its own, in a manner of speaking. Further extensions of forms leads eventually to highly abstract, symbolic (tertiary) systems of representation. The PMS is the "default" system, while the SMS and TMS respectively are extensional systems (see figure 55 below).

The *dimensionality principle* asserts that the PMS, the SMS, and the TMS are interactive systems in human knowing and knowledge-making: i.e., when one system cannot be applied to a situation another one can usually be used in its place. This principle thus provides a framework for showing an interrelation and interdependence among all areas of knowledge, from language to science and mathematics. A methodological corollary of this principle is that, rather than examining signs, texts, codes, and metaphors as isolated phenomena, SA would see each of these as types of models (singularized, composite, cohesive, connective) and thus as interconnected to each other. This corollary is known as the *interconnectedness principle*. MST is thus intended as a unifying theory of semiotics, as well as a framework for conducting a cross-species study of semiosis. The *structuralist principle*, finally, asserts that certain elemental structural properties characterize all models: i.e., they beget their forms and meanings by paradigmaticity, syntagmaticity, synchronicity, diachronicity, and signification. *Paradigmaticity* is a differentiation property of forms. The feature that yields a difference in form can be unraveled by comparing a specific form with another form taken from the same system. This comparison, known technically as *opposition*, will reveal which minimal feature in a form is the critical distinctive one. *Syntagmaticity* is a combinatory property of forms. Simply put, forms are recognizable as belonging to certain systems by virtue of the fact that their constituent elements are put together in patterned ways. *Synchronicity* refers to the fact that models assume a specific form at a particular point in time; while *diachronicity* refers to the fact that forms are subject to change over time. Finally, *signification* is the relation that holds between a form and its referent (or referential domain). For example, in the case of a primary model, the relation is that of simulation producing a denotatum; in the case of a secondary model, it is that of extension, producing a connotatum, or that of indication; and in the case of a tertiary model it is that of further extension, producing an abstract, symbolic form.

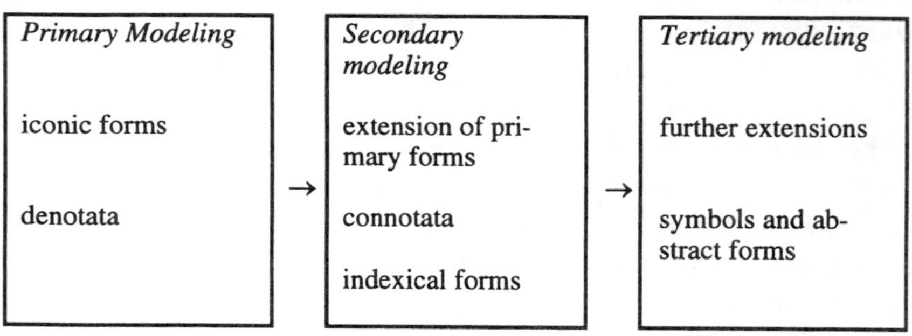

Figure 55. Modeling extensionality

The phenomena that SA would target are summarized in table 8 below:

Table 8. Phenomena targeted by SA

Phenomena	Manifestations
Singularized Modeling	individual signs (signals, symptoms, icons, indexes, symbols, names)
Composite Modeling	verbal and nonverbal texts
Cohesive Modeling	verbal and nonverbal codes
Connective Modeling	metaforms, meta-metaforms, meta-symbols
Primary Modeling	simulation, sense-inference, iconicity
Secondary Modeling	extension of primary models, indexicality
Tertiary Modeling	further extension, symbolicity
Representation	any externalized form
Dimensionality	pattern of interdependence among a form and its extensions
Extensionality	produces connotata and symbols
Interconnectedness	primary, secondary, and tertiary forms are interconnected, i.e., one presupposes the other two
Paradigmaticity	features that keep forms distinct
Syntagmaticity	combinatory structure of forms
Signification	denotata, connotata, symbols

SA can now be formalized into a series of methodological tasks, which are shown in figure 56 below in sequential order. These tasks consist essentially in:

- determining the type of model that a form manifests in a specific instantiation or application (singularized, composite, cohesive, connective);
- determining the modeling system from which it emanates in terms of its instantiation or application;
- determining the representational processes involved in its instantiation or application (representationality, dimensionality, etc.);
- determining the ways in which its structural properties (paradigmaticity, syntagmaticity, etc.) manifest them-

selves in the instantiation or application under investigation.

Task 1: Determining the Type of Model

↓

Singularized
Composite
Cohesive
Connective

↓

Task 2: Determining the Modeling System from Which it Emanates

↓

PMS
SMS
TMS

↓

Task 3: Determining the Representational Processes Involved

↓

Form and Function of the Representation (Representational Principle)

Interaction of the Systems in Construction (Dimensionality Principle)

Extensional Processes (Extensional Principle)

Relation of the Different Forms of a Model to Each Other
(Interconnectedness Principle)

↓

Task 4: Determining the Structural Properties of the Model

↓

Paradigmaticity
Syntagmaticity
Synchronicity
Diachronicity
Signification

Figure 56. The methodological tasks involved in SA

This shows how SA would take systematically into account the various facets of semiotic analysis in an integrative fashion. As we have seen throughout this book, organisms possess species-specific modeling capacities that allow them to respond in kind to their outer experiences and needs. Verbal modeling is unique to the human species, and thus requires special kinds of analytical procedures. All other modeling systems in Nature are nonverbal. Language is verbal, but not necessarily vocal (e.g., it can be communicated also by means of alphabet characters, gestures, etc.); speech, on the other hand, is both vocal and verbal. Communication systems are formed in the organism by exposure to appropriate input in social context and are subject to change or even dissolution over time. In all species, other than the human, systems are formed primarily through the biological channel; human beings acquire their ability to communicate both from biology and from culture.

Once the nature of the modeling process has been ascertained, then its forms and functions can be deduced or inferred from observation of the semiosic behavior involved. Thus, the cross-species investigative focus of SA has clear implications for ethology and animal psychology, as well as for traditional semiotic theories and branches of semiotic science. Its central proposal in the study of anthroposemiosis, and thus its methodological bias, is that the tendency in human representation is to produce, first and foremost, a sensory model of some referent or referential domain and then, by extensional processes, to make it encompass increasingly larger domains of meaning. This "flow" from iconicity to connotativity and symbolicity, i.e., from concrete, sensory modes of representation (and knowing) to complex, abstract modes, characterizes most of human modeling (and knowledge-making).

5.2 Anthroposemiosis

The uniqueness of anthroposemiosis lies in the fact that the human species is endowed with three modeling systems that work interdependently and interactively in the production of models and, thus, of knowledge. But as SA reveals, the uniqueness of anthroposemiosis is nevertheless grounded in the biology of the human species.

5.2.1 Evolutionary antecedents

There are four distinctive events that occurred in human evolution which have shaped the emergence of this three-tiered modeling system–namely, bipedalism, brain growth, tool-making, and tribalism. One of the earliest of the major hominid characteristics to have evolved, distinguishing the species *Homo* from its nearest primate relatives–the gorilla, chimpanzee, and orangutan–was an adaptation to a completely erect posture and a two-footed striding walk. Almost all other mammals stand, walk, and/or run on four limbs. Those that stand on two have quite different postures and gaits from humans–kangaroos hop on their two feet; some monkeys may only on occasion walk bipedally, especially when carrying food; chimpanzees are capable of brief bipedal walks, but their usual means of locomotion is knuckle-walking, standing on their hind legs but stooping forward, resting their hands on the knuckles rather than on the palms or fingers. So, even though forms of bipedalism are observable in other primates, they are unlike the human type: all other forms of bipedal walking involve straight or bowed spines, bent knees, grasping feet, and some use of the hands to bear part of the body weight during locomotion. The uniquely S-shaped spinal column of humans places the center of gravity of the body directly over the area of support provided by the feet, thus ensuring stability and balance in the upright position.

Fossils discovered in Africa provide evidence that hominids walked erect and had bipedal stride even before the great increase in brain size. Complete bipedalism freed the human hand, allowing it to become a supremely sensitive limb for precise manipulation and grasping. The most important structural detail in this refinement was the elongated human thumb, which could rotate freely for the first time and, thus, be fully opposable to the other fingers. No doubt, this development made tool-making and tool-use possible. Moreover, many linguists claim that the erect posture gave rise to the subsequent evolution of the physiological apparatus for speech, since it brought about the lowering and positioning of the larynx for controlled breathing. In a phrase, bipedalism, tool-making, brain growth, and language were probably intertwined in human evolution.

Although other species, including some non-primate ones, are capable of tool use, only in the human species did complete bipedalism free the hand sufficiently to allow it to become a supremely sensitive and precise manipulator and grasper, thus permitting proficient tool-making and tool-use in the species. The earliest stone tools are nearly 2.5 million years old. One and half million year old sites in various parts of eastern Africa have been found to contain not only many stone tools, but also animal bones with scratch marks that could only have been left by human-like cutting actions. One thing is for certain–only in the human species does one find the capacity to fashion a great diversity of tools from the raw materials found in the environment to meet virtually any need that may arise.

From early on in their evolution, humans have made tools in order to model, understand, and modify the world. This explains why modern humans think of objects–jewelry, clothes, furniture, ornaments, tools, toys, etc.–as valuable artifacts. An *artifact* is, literally, "something made" with a specific cultural function in mind. In a fundamental sense, human cultures are museums of the artifacts that have been produced by the sapient animal.

Technology is the general term used for describing the processes by which human beings fashion objects in order to increase their understanding of, and control over, the material environment. The term is derived from the Greek words *tekhne,* which refers to an 'art' or 'craft', and *logia,* meaning an 'area of study'. Many historians of science argue that technology has not only become an essential condition of advanced, industrial civilization, but also that the rate of technological change has developed its own momentum in recent centuries. Innovations now seem to appear at a rate that increases geometrically, without respect to geographical limits or social systems. These innovations tend to transform traditional signifying orders, frequently with unexpected social consequences.

Like most other species, humans have always lived in groups. Group life enhances survivability by providing protection, shelter, and other safeguards against both enemies and abrupt changes in the surroundings. But at some point in their evolutionary history–probably around 100,000 years ago–bipedal hominids had become so adept at tool-making, communicating, and thinking in symbols that they became *consciously* aware of the advantages of a group life based on a signifying order. By around 30,000 to 40,000 years ago, the archeological evidence suggests that human groups became increasingly characterized by communal customs, language, and the transmission of technological knowledge through a signifying order. Anthropologists have labeled these forms of group life *tribal*.

The *tribe* remains the type of collectivity to which human beings instinctively relate even in modern times. In complex city-societies, where various cultures, subcultures, countercultures, and parallel cultures exist in constant competition with each other, where the shared territory is so large that it becomes an abstraction or figment of mind, the tendency for individuals to relate to tribal-type groupings or arrangements that exist within the larger societal context manifests itself regularly and predictably. People continue to perceive their membership in smaller groups as more directly meaningful to their lives than allegiance to the larger society or nation. This inclination towards *tribalism*, as Marshall McLuhan emphasized throughout his career, reverberates constantly within modern-day humans, and may be the source of the angst and sense of alienation that many city-dwelling individuals feel, living in large, impersonal social systems.

In sum, bipedalism, tool-making, and tribalism–all dependent on considerable brain growth–are the evolutionary antecedents of the three modeling systems of anthroposemiosis. The coherence of human-made forms into an overarching system of meaning, which is culture, is psychologically and socially unique among species.

5.2.2 Sense-inference

The starting point in studying anthroposemiosis from the standpoint of SA is to look for what we have called *sense-inference* in the origin of all human representational activities. The *Sense-Inference Hypothesis* (SIH) (chapter II, §2.1.1) posits that all model-making efforts in infancy and childhood are initially grounded in the experiential realm of the senses. As suggested earlier, this initial stage of knowing can be called a *cognizing* stage, while the stage when the sensory units of knowing are transformed into representational forms can be called the *recognizing* stage. The SIH implies, as a corollary, that the iconic mode of representation is the primary means by which *recognition* is made possible in early forms of representation. The simulative, replicational, or imitative forms constructed and expressed through verbal and nonverbal channels constitute the primary models through which children attempt to come to grips with the outer world and their internal consciousness.

But sense-inference is not limited to childhood. It continues to operate in daily life as needed. Suppose someone encounters an unknown object. Lacking any pre-existing form for referring to it, the person will explore its sensory qualities: i.e., what it looks like, what it feels like, etc. Now, if the object is deemed to have some value or use, then the individual will assign it a name. The way in which s/he will do so is rarely arbitrary. The person will typically attempt to capture some sensory property of the object and then encode it in the name. Let us assume that the object had a very pronounced *smoothness* to the touch; then the person might come up with a name that is both alliterative and analogical (vis-à-vis the word *smooth*). One possible candidate is *smoor*. As time passes, and the usefulness of *smoors* becomes obvious to other people, then extensional processes intervene to assign connotative and symbolic meanings to the word *smoor*: e.g., a *smoor* can be applied to connote a certain type of *smoorness* in social life, or as a meta-symbol for human personality: [people = smoors].

It is important to note the following structural aspects of the above modeling process:

- The word was acceptable to those who came to use it because it was well-formed paradigmatically and syntagmatically: a word such as *spmoor*, on the other, would not have been even contemplated by the form-maker since it would violate the structural properties of English words.
- The word had an initial denotatum [smoothness] that was extended by the secondary and tertiary modeling processes into connotata and symbols.
- As a singularized form it can be incorporated into composite, cohesive, and connective modeling systems to encode complex referents.

The use of SA to explain human knowledge makes it clear that novel representational tasks (*cognizing events*) are more apt to be more comprehensible if they are represented iconically at first. Take, as an example, how three well-known atomic theories in physics are, in fact, interconnected to each other in terms of the notions discussed in this chapter (and this book): (1) the *Rutherford Model* which portrays the atom space as a tiny solar system; (2) the *Bohr Model,* which adds 'quantized' orbits to the *Rutherford Model*; and (3) the *Schrödinger Model*, which posits the idea that electrons occupy regions of space.

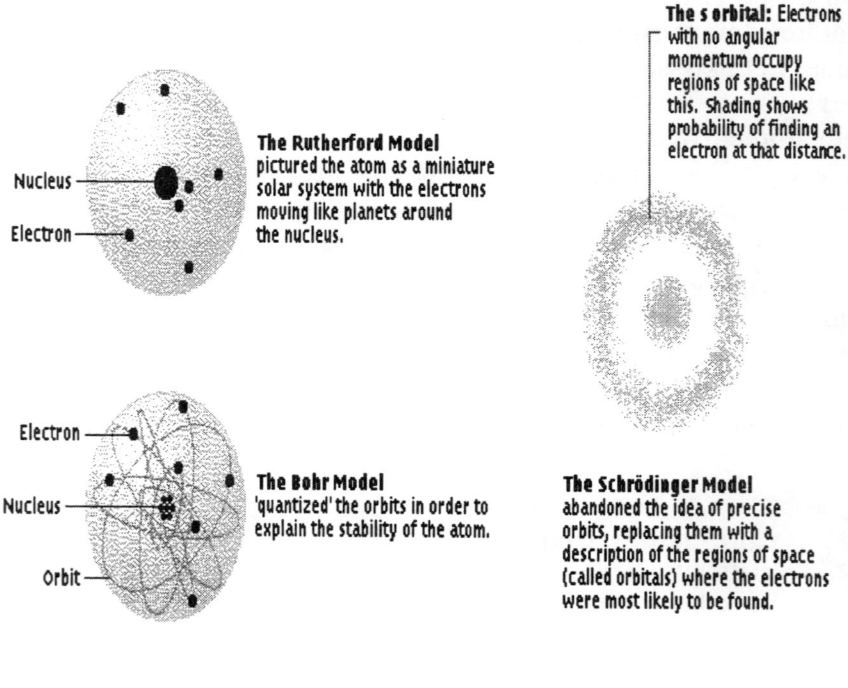

Figure 57. Different models of atomic structure

The way in which each model is composed is hardly haphazard: each one attempts to show atomic structure according to specific types of experimental data. The referential domain is the same in all three cases, namely [atomic structure]; but each diagram provides, literally, a different "mental view" of the same domain–a domain that is not directly accessible to vision. The *Bohr Model* is, in effect, an extension of the *Rutherford* one, and the *Schrödinger Model* an extension of the previous two. The model envisioned by Rutherford in which electrons move around a tightly-packed, positively-charged nucleus, successfully explained the results of scattering experiments, but was unable to explain atomic emission (why atoms emit only certain wavelengths of light). Bohr began with Rutherford's model, but then postulated further that electrons can only move in certain quantized orbits. His model was thus able to explain certain qualities of emission for hydrogen, but failed for other elements. Schrödinger's model, in which electrons are described not by the paths they take but by the regions where they are most likely to be found, can explain certain qualities of emission spectra for all elements.

5.3 Zoosemiosis

As we have shown throughout this book, SA extends its purview to include the documentation of analysis of phytosemiotic and zoosemiotic modeling phenomena in a comparative way. Attempts at classifying animal behaviors go as far back as 400 BC. But it was Aristotle who devised the first system of classifying animals, when he arranged groups of animals according to mode of reproduction and habitat.

One of the more influential figures in the history of physiology, the Greek physician Galen, dissected farm animals, monkeys, and other mammals and described many features accurately, although some were wrongly applied to the human body. His misconceptions, especially with regard to the movement of blood, remained virtually unnoticed for hundreds of years. Until the Middle Ages, the study of animals was shaped by a concoction of superstitions about animals. Indeed, only in the twelfth century did it begin to reemerge as a science, after the German scholar St. Albertus Magnus (c. 1200-1280) revived the ideas of Aristotle. Subsequently, Leonardo da Vinci's (1452-1519) dissections and comparisons of the anatomical structure of humans and other animals led the Belgian physician Andreas Vesalius (1514-1564) to circulate da Vinci's writings and thus to establish the principles of comparative anatomy.

Classification dominated the study of animals throughout most of the seventeenth and eighteenth centuries. The Swedish botanist Carolus Linnaeus (1707-1778) developed a system of nomenclature that is still used today, and established taxonomy as a discipline. Other leading taxonomists of the era were the French biologists Comte Georges Leclerc de Buffon (1707-1788) and Georges Cuvier (1769-1832). In 1839 two Germans, Matthias Schleiden (1804-1881) and Theodor Schwann (1810-1882), proved that the cell is the common structural unit of all living things, a concept that provided the scientific basis for the development by the Frenchman, Claude Bernard (1813-1878), of the study of animal physiology.

Expeditions in the eighteenth and nineteenth centuries gave animal scientists the opportunity to study plant and animal life throughout the world. The most famous expedition was the voyage of the Beagle in the early 1830s, during which Charles Darwin (1809-1882) classified the plant and animal life of South America and Australia and developed his theory of evolution by natural selection. Shortly thereafter, the Austrian monk Gregor Mendel (1822-1884) formulated the concept of particulate hereditary factors–later called *genes*. In the twentieth century, the classification of animals has become quite sophisticated, broadening its range to include notions taken from such disciplines as genetics, ecology (the study of the interactions between animals and their environment), and biochemistry.

Zoosemiotics concentrates on the different divisions of animal life according to modeling system. By investigating the classification, distribution, life cycle, and evolutionary history of the particular animal or group of animals under study, zoosemiotics studies how physical phenomena such as the morphology (animal structure), histology (body tissue system), and cytology (cells and their components) of an animal species relates to the manifestations of modeling in that species. Among the kinds of background information that is necessary in order to conduct a zoosemiotic SA, the following figure prominently:

- the description of an animal species;
- the developmental history of that species;

- the territorial distribution of the species;
- the collection of data on the species' communicative behavior as it manifests itself in camouflage, courtship, mating, etc.

5.3.1 Comparative perspective

SA is, by its very nature, comparative. Animals have body structures that differ significantly. These constitute the source of differences in modeling systems. To construct explanatory schemes of modeling behavior, SA must thus examine and compare the anatomy, biochemistry, genetic systems, behavioral systems, and fossil histories of as many organisms as possible. More than 1.5 million different groups have been identified and at least partly described, and many more remain to be studied. As von Uexküll (1909) perceptively noted, the two disciplines–biology and semiotics–overlap considerably; the former focusing more on uncovering evolutionary relations in the physical structure of animals, the latter on how anatomical and morphological structure constrain modeling activities.

The three Nobel Prize-winning founders of ethology–Konrad Lorenz (1903-1989) of Austria, Nikolaas Tinbergen (1907-1988) of the Netherlands, and Karl von Frisch (1886-1982) of Germany–uncovered four basic ways in which animals behave across species. Needless to say, these are important in any zoosemiotic application of SA:

Sign stimuli (releasers)

These are cues that enable animals to recognize critical referents when they encounter them for the first time. Baby herring gulls, for example, must know from the outset to whom they should direct their begging calls and pecks in order to be fed. The young chick's recognition of a parent is based entirely on the sign stimulus of the vertical line of the parent's bill and a red spot moving horizontally. A wooden model of the bill works as well as the real parent. The most widespread uses of sign stimuli in the animal world are in communication, hunting, and predator avoidance. The young of most species of snake-hunting birds, for instance, innately recognize and avoid deadly coral snakes; young fowl and ducklings are born able to recognize and flee from the silhouette of hawks; the bee-hunting wasp recognizes honeybees by means of a series of releasers; and the list could go on and on.

Motor Programs

These are self-contained circuits able to direct the coordinated movements of many different muscles to accomplish a task. The dancing of sticklebacks, the stinging action of wasps, and the pecking of gull chicks are all manifestations of motor programming. In the human species, walking, swimming, bicycle riding, and shoe tying, for example, begin as laborious efforts requiring full, conscious attention. After a time, however, these activities become so automatic that, like innate motor programs, they can be performed unconsciously and without normal feedback. This need for feedback in only the early stages of learning is widespread. Both songbirds and humans, for example, must hear themselves as they begin to vocalize.

Drive

This is the innate impulse informing animals when to migrate, when (and how) to court one another, when to feed their young, and so on. Geese, for example, will only roll eggs from about a week before egg laying until a week after the young have hatched. At other times, eggs have no meaning to them. In birds, preparations for spring migration, territorial defense, and courtship behavior are activated by the period of daylight. This alters hormone levels in the blood, thereby producing changes in behavior.

Programmed Learning

This is the ability of a species to learn only those things that are relevant to its life needs. One famous example of programmed learning is *imprinting*. The young of certain species must be able to follow their parents almost from birth. Each young animal must quickly learn to distinguish its own particular parents from all other adults. It has been found that imprinting occurs: (1) during a specific time, or critical period; (2) in a specific context, usually defined by the presence of a sign stimulus; (3) in such a way that an animal remembers only a specific cue and ignores others; (4) in the absence of rewards. In a world full of stimuli, imprinting enables an animal to know what to learn and what to ignore.

A systematic analysis of these four areas of comparison will usually indicate what modeling phenomena are, in fact, comparable, and which ones are unique. Take, as an example, the notion of *fetish*, which hitherto has been considered a unique feature of human behavior.

The English word *fetish* was directly adopted from the Portuguese substantive *feitiço* 'charm, sorcery' (Spanish *hechizo*, both from the Latin *facticius*, 'factitious', meaning 'artificial, skillfully contrived'). Originally, the term was applied to any of the objects used by the people of the Guinea coast and neighboring regions as talismans or amulets. Portuguese sailors allegedly coined the term in the fifteenth century when they observed the veneration that West Coast Africans had for such objects, which they wore on their bodies (Herskovits 1947: 368). The earliest English citation dates from a 1613 work by Purchas, titled *Pilgrimage* (VI, xv, 651): "Hereon were set many strawen Rings called *fatissos* or *Gods*".

Historians, following Brosses (1760), began using *fetish* in the wider sense of an inanimate object worshipped on account of its supposed inherent magical powers in the eighteenth and nineteenth centuries. In 1869, McLennan, who framed *totemism* as an anthropological concept, also invented the notorious formula: totemism is fetishism plus exogamy and matrilineal descent (Lévi-Strauss 1962: 18).

The divine in many religions is chiefly sought in objects, which are thought of as capable of capturing natural forces–inanimate things, such as pieces of wood, relics of saints, statues, crosses; food and drink, such as bread and wine or baptismal water; living things, such as the totemic animal of the group, the sacred cow, the sacred tree; processes, such as the movements of sacred dancing; and so on. In primitive forms of religion, when the object itself is perceived to be divine, that object is designated a *fetish*. This is thought typically to have the power to heal or cure sickness, to influence social relations, and even to induce erotic desire.

Clearly, it is the latter perception that led to the eventual espousal of the term in clinical discourse to describe the enhancement of sexual desire in the presence of a type of object *Psychopathia sexualis* (1886), by the forensic psychiatrist Richard von Krafft-Ebing (1840-1902), contained the first systematic collection of data relating to "pathological" fetishism. This text, with its view of sex as perverted and disgusting, came to exert a great and seemingly perpetual influence on the Western world. Krafft-Ebing wrote extensively of sex crimes and sexual variations or deviations, which he considered based upon genetic defects. It was, in fact, probably Krafft-Ebing who first referred to the notion of *fetishism* as a "perversion", and thus as something that required social sanctions to control it. According to his description, a fetish was a nonhuman object–a part of the body or something contiguous to it, such as clothing–which served as an impetus to sexual arousal and orgasm. He considered all acts other than marital coitus for the purpose of procreation, and all surrogates for coitus, as reprehensible perversions.

A textbook of *Psychiatry* by Freedman, Kaplan, and Sadock written in 1972 likewise defined fetishism in an explicitly sexual way: "The process of achieving sexual excitement and gratification by substituting an inanimate object such as a shoe, piece of underwear, or other article of clothing for a human love object". Ever since, it has been very common in the therapeutic literature to find references to the attraction a patient may have for an inanimate object as "inordinate" or "pathological".

There is also the widely-held notion among clinicians that fetishism is much more common in men than in women. This presumption was initiated by Freud (see Vigener 1989) and reinforced by Kinsey (1953: 679) and his collaborators who considered fetishism to be an "almost exclusively male phenomenon". Freud too held that fetishism was *the* male perversion par excellence. Schor (1985: 303) put it in a nutshell as follows: "female fetishism is, in the rhetoric of psychoanalysis, an oxymoron".

The most extensive recent study of the fetish in sexology is to be found in John Money's *Lovemaps* (1986: 261), in which he offers a conventional definition: "an object or charm endowed with magical or supernatural power; an object or part of the body charged, for a particular person, with special sexual-erotic power". However,

Money (1986: 265) also points out that there "is no technical term for the reciprocal paraphiliac condition in which the fetish, for example, a uniform, must belong to the self". Money classifies (1986: 65) tangible objects in addition to those appealing to the eye, as either haptic or olfactory, available to immediate perception or in fantasy. Although Money does not emphasize it, he acknowledges that the use of fetishes by females is considerably more prevalent than has been explicitly recognized in the literature.

Children of both sexes frequently cling to an object. Such an object may be related by contiguity to a parent or to the infant's early material surroundings. According to some psychiatrists (Freedman et al. 1972: 637), this "is a security operation that should be distinguished from fetishism in which the normal sexual object is substituted by another" which, as these authors emphasize "is not known to occur in childhood". However, this judgment may be due to the prejudice in psychiatry that a fetish, in order to be defined as such, must produce genital sexual satisfaction, and thus that the use of objects to produce a fetishistic effect necessarily occurs relatively late in adolescence (Sperling 1963; Roiphe 1973; Bemporad et al. 1976).

Now, whereas psychiatrists consider fetishism to be a unique human trait, the broader purview of SA shows this to be patently unwarranted. A fetish is a form that gains its effectiveness as a releaser. According to Guthrie's (1976: 19) excellent account of the anatomy of social organs and behavior, *releaser* signs "occur in the form of extra-large social organs, i.e., increasing signal strength by increasing signal amplitude". Thus, in certain species of animals, antlers and horns function as fetishistic estimations of rank; this is why they "grow to gigantic size among the older males, or develop specialized modifications, like filling in between the tines to form palms, thereby increasing the visual effect from a distance".

In particular, anal and genital organs–about which humankind harbors so many taboos–tend to become fetishistic forms for several reasons: in part, because mammals, having, in general, a well developed smelling apparatus, tend to use feces and urine as a part of their signaling behavior, and, in part, because of the sexual overtones of different mammalian ways of urination.

The phenomenon of the releaser stimulus has been demonstrated many times in studies of animal behavior, especially in one exemplary piece of work by Tinbergen and Perdeck (1950). In brief, these two investigators found that they could devise a releaser stimulus consisting of an artificial model in which some sign aspects are exaggerated relative to the natural object. Such a stimulus was provided by a long red knitting needle with three white rings near the tip. This was more effective than a naturalistic head and bill of an adult gull in evoking a pecking response from herring-gull chicks.

Writing about domestic cats, the ethologist Leyhausen (1967) observed that "substitute objects" can become supernormal objects, as when a cat disports itself with a ball of paper in an intensive catching game, while perfectly "adequate" prey mice run around under its very nose. Indeed, fetishistic attachments are commonplace among vertebrates–particularly in mammals, as well as in many birds.

As SA demonstrates, the definition of fetish requires a considerable revision in psychiatry. In this perspective, what in the sexological literature is called a fetishistic attachment may be viewed as a form of mal-imprinting. As Morris (1969: 169) writes: "Most of us develop a primary pair-bond with a member of the opposite sex, rather than with fur gloves or leather boots...but the fetishist, firmly imprinted with his unusual sexual object, tends to remain silent on the subject of his strange attachment...The fetishist...becomes isolated by his own, highly specialized form of sexual imprinting".

Fetishism in the human species is a semiosic phenomenon that saliently illustrates how semiosis interconnects biological, psychological, and cultural processes. In a fundamental sense, culture is a repository of the fetishistic artifacts that people make. The process by which this system takes on meanings can be called *objectification*. This term is meant to characterize the fact that objects are interconnected to each other in terms of the signifying order. As Barthes (1970: 23) put it, in the same way that words issue forth from the "infinite sea of the already written", so too the objects people make can similarly be said to issue forth from "the infinite sea of the already made". Objects are extensions of the species and, in turn, influence its future evolution. Since our bipedal origins, we have made tools, but as a consequence, we have had to live with the fact that our tools have thereafter shaped us. So, in a fundamental sense, all tools are fetishes, to one degree or other.

5.3.2 Interspecies communication

The question of whether it is possible to teach animals human representational systems, or even whether interspecies communication, without human intervention, is realizable falls outside the purview of SA. It is nevertheless an issue that cannot be ignored totally here, given the ongoing experiments with teaching simians various forms of human language. So, we will look briefly at this final issue here.

While we may not be able to communicate with other species in the same ways that we do with each other, there is a level at which we do indeed "make contact" with some species. There is no doubt that a household cat and a human enter into a rudimentary form of communication on a daily basis. Sharing the same living space, and being codependent on each other for affective exchanges, people and cats do indeed communicate feeling-states to each other. The two species do this by signaling exchanges and by utilizing modes of communication other than those that are based on species-specific forms. Tones, postures, movements are the signifiers of a mutually-developed "interspecies communication code". This code taps basic sensory *modes* and transmission *media* that cut across human and feline communication systems, emerging "adaptively" from shared experiences.

What makes the human system of communication unique among species, and thus unlikely to be transferable to other species, is that it can take place through more than one *mode* and *medium* of transmission. As we have seen in this book, human communication utilizes all the sensory modes of communication (visual, tactile, etc.), as well as purely verbal ones. The latter modes are unique among species. The types of choices of transmission *media* through which human communication unfolds are also highly versatile. There are three kinds of media through which humans communicate:

- *natural media*, such as the voice (speech), the face (expressions), and the body (gesture, posture, etc.);
- *artifactual media*, such as books, paintings, sculptures, letters, etc.;

- *mechanical media*, such as telephones, radios, television sets, computers, videos, etc.

A verbal message, for instance, can be encoded for *natural* transmission, if it is articulated with the vocal organs; or else it can be encoded in an *artifactual* way, by means of markings on a piece of paper through the medium of writing; and it can also be converted into radio or television signals for *mechanical* (electromagnetic) transmission. Needless to say, artifactual and mechanical media are specific to human communication. There also appears to be a cause-and-effect relation between the kinds of media developed and human evolution. Babylon was transformed by the invention of cuneiform writing, Egypt by written hieroglyphics, the ancient Greeks by the alphabet, and so on. The paper and the printing press brought about new ways of thinking across Europe and paved the way for the European Renaissance and the Protestant Reformation. And, of course, electronic forms of communication (radio, television, e-mail, etc.) are having just as much impact, if not more, on modern society as the invention of writing had on ancient cultures, and printing on medieval people.

Communication is an outgrowth of the necessity to engage in joint action. The exchange of signals help animals find food, migrate, or reproduce. But while other animals use a limited range of signals to communicate, humans have developed complex modes and media of communication that are used not only to ensure survival, but also for the expression of ideas and emotions, for the telling of stories, for the recollection of the past, and for the negotiation of meaning with one another. These types of functions of communication are unknown in other species. In all other species communication has a fairly straightforward function: to exchange information about the environment in order to ensure survival. This is why honeybees dance in specific patterns that tell other members of the hive where to find food, why insects regularly use pheromones to attract mates, why elephants emit low-pitched sounds to get the attention of other members of the herd over many miles, why chimpanzees use facial expressions and body language to express dominance or affection with each other, why whales and dolphins make vocal clicks, squeals, or sing songs to exchange information about feeding and migration, and to locate each other.

5.4 Concluding remarks

As Peirce (I: 538) cogently argued, "Every thought is a sign". But, as he also wrote, "Not only is thought in the organic world, but it develops there" (V: 551). This statement encapsulates why modeling is characteristic not only of the human world, but of the entire organic world, where, indeed, it developed. The *Umwelt* and *Innenwelt* of all animals, as well as the feedback links between the two, are created and sustained by the particular biology that characterizes a species. A model is a semiosic production with species-specific biological features for its utilization.

This is as true of bees (Peirce V: 551) as it is, on a far vaster scale, of Isaac Newton's and Albert Einstein's grand models of the universe. Einstein constructed his model out of nonverbal signs, laboring long and hard to transmute his creation into conventional words, so that he could communicate it to others. "The words or the language, as they are written or spoken", Einstein wrote in a letter to Hadamard (1945: 142-143), "do not seem to play any role in my mechanism of thought. The psychical entities which seem to serve as elements in thought are certain signs and more or less clear images which can be 'voluntarily' reproduced and combined".

As we have written throughout this book, the relatively simple, nonverbal models that animals produce are natural forms which must fit "reality" sufficiently to secure the survival and "sanity" of the members of a species in their ecological niche. In human beings, the modeling instinct is so pervasive and powerful that it often becomes very sophisticated indeed in the adult life of some individuals, as borne out by Einstein's testimonial, or by what we know about Mozart's or Picasso's ability to model intricate auditory or visual referents in their heads in anticipation of transcribing them onto paper or canvas. Language, metaforms, meta-symbols, as far as we know, are unique to anthroposemiosis. These make it possible for humans not only to represent immediate reality, but also to frame an indefinite number of possible worlds. The modeling capacity in humans has led to what Bonner (1980: 186) calls "true culture", requiring "a system of representing all the subtleties of language", in contrast to "nonhuman culture". It is on this level, defined as tertiary, that nonverbal and verbal sign assemblages blend together in the most creative modeling system that Nature has thus far produced.

The main purpose of this book has been to show that SA can be used to provide a synthetic framework for relating what would appear to be disparate and heterogeneous findings on semiosic phenomena to each other. The framework developed by SA forces us to see ourselves as partners with all other life forms in the natural scheme of things, even though our particular brand of modeling has led us to live primary through the very symbolic forms that it allows us to produce. Cassirer (1944: 25) has put it eloquently in the following words:

> No longer in a merely physical universe, man lives in a symbolic universe. Language, myth, art, and religion are parts of this universe. They are varied threads which weave the symbolic net, the tangled web of human experience...No longer can man confront reality immediately; he cannot see it, as it were, face to face. Physical reality seems to recede in proportion as man's symbolic activity advances. Instead of dealing with the things themselves man is in a sense constantly conversing with himself. He has so enveloped himself in linguistic forms, in artistic images, in mythical symbols or religious rites that he cannot see or know anything except by the interposition of this artificial medium.

We conclude by pointing out that human beings have always shown the ability to disentangle themselves from the "tangled web", as Cassirer called it. In each individual human being there is a continual juxtaposition of individually-based vs. culturally-based modeling. Indeed, culture cannot stamp out the individual human being's need and quest for new forms of meaning. The ability to make signs anew to represent changing realities, new ideas, new ways of thinking is the essence of anthroposemiosis. This innate creative propensity is the reason why cultural symbols are constantly being modified to meet new demands, new ideas, new discoveries, new challenges.

Human modeling often involves things that cannot be seen. Unique among species, connective modeling allows humans to get a look, so to speak, at this hidden world. The poet and the scientist alike use similar modeling systems to extrapolate a suspected inner connection among things. When their models are accepted as fact, they enter human life, taking on an independent conceptual existence in the real world, and thus can suggest ways in which to bring about changes to the world.

As we stated in the preface to this book, model-making typifies all aspects of human cognitive and social life. From toy and miniature models to scientific theories of the universe, models are so common that we hardly ever take notice of their importance and of their raison d'être in our species. Model-making constitutes a truly astonishing evolutionary attainment, without which it would be virtually impossible for us to carry out our daily life routines and to encode knowledge. We reiterate that the presence of a modeling instinct in the human species is to human intellectual and social life what the physical instincts are to its biological life. As Thomas Szasz (1920-), the great American psychiatrist aptly put it, "in the animal kingdom, the rule is, eat or be eaten; in the human kingdom, define or be defined".

Glossary of technical terms

A

abduction	process by which a new concept is formed on the basis of an existing concept which is perceived as having something in common with it
abstract concept	a mental form whose external referent cannot be demonstrated or observed directly
actant	a unit of narration (a hero, an opponent) that occurs in all kinds of stories
adaptor	bodily movement indicating or satisfying some emotional state or need: e.g. scratching one's head when puzzled; rubbing one's forehead when worried
affect displays	hand movements and facial expressions communicating emotional meaning
alliteration	the repetition of the initial consonant sounds or features of words
alphabet	graphic code providing individual characters that stand for individual sounds (or sound combinations)
analogy	equivalence property of forms, by which one type of form can be replaced by another that is perceived as being comparable to it
annotation	interpolation or assignment of a subjective and/or social meaning to a form (sign, text, etc.)
annotatum	specific personal meaning that a form elicits
anthroposemiosis	semiosis in humans
anthroposemiotics	the study of semiosis, modeling, and representation in humans
artifact	an object produced or shaped by human craft, especially a tool, a weapon, or an ornament of archaeological or historical interest
artifactual media	media such as books, paintings, sculptures, letters, etc. made by human beings in order to transmit messages

artificial model	a model produced artificially, i.e. intentionally, by a human being

B

basic concepts	concept having a typological function
binary iconic feature	a sensory property inferred in a referent (e.g. [upward motion]
binary opposition	minimal difference between two forms
biosemiotics	branch of semiotics aiming to study semiosis, modeling, and representation in all life forms
bipedalism	walking upright on two feet

C

channel	the physical means by which a signal or message is transmitted
character	alphabetic symbol
code	system of signifying elements which can be deployed to represent types of phenomena in specific ways
coevolution	the sociobiological theory that genes and culture are evolving in tandem
cognitive style	the particular way in which information and knowledge are processed
cognizing stage	stage whereby a referent is modeled in a sensory way
cohesive modeling	modeling strategy that is inclusive of referents that are deemed to share features
communication	capacity to participate with other organisms in the reception and processing of specific kinds of signals
composite modeling	modeling strategy encoding complex or composite referents or referential domains in a composite manner
concept	mental form
conceptual metaphor	generalized metaphorical formula that defines a specific abstraction (*love is a sweet taste*)

concrete concept	mental form whose external referent is demonstrable and observable in a direct way
connective form	form which results from the linkage of different types of referents (or referential domains)
connotation	extension of a form over a new meaning domain that is recognized as entailing the features of the form by implication
connotative extensional modeling	the process of extending the meanings of primary forms to encompass connotative meanings
connotatum	extended meaning of a form
context	situation—physical, psychological, and social—in which a form is used or occurs, or to which it refers
conventional sign	sign that has no apparent connection to any perceivable feature of its referent
cultural modeling	the association of various source domains with one target domain, producing an overall, or culture-specific model, of the target domain
culture	the system of daily living that is held together by a signifying order (signs, codes, texts, connective forms)
cuneiform writing	system of characters formed by the arrangement of small wedge-shaped elements and used in ancient Sumerian, Akkadian, Assyrian, Babylonian, and Persian writing

D

decoding	use of a code to decipher forms
deduction	process of applying a mental form to a specific referent
deixis	process of referring to something by pointing it out or specifying it in some way
denotation	initial, or intensional, meaning captured by a form
denotatum	the initial signified of a sign
diachronicity	change in a form over time

dimensionality principle	principle asserting that the PMS, the SMS, and the TMS are interactive systems in human knowing: i.e. when one system cannot be applied to a situation another one can usually be used in its place
distinctive feature	minimal element that makes up a form and which singularly or in combination with other distinctive features serves to differentiate its meaning from other forms
drive	the innate impulse informing animals when to migrate, when (and how) to court one another, when to feed their young, and so on

E

emblems	gesture that directly translate words or phrases: e.g. the *Okay* sign, the *Come here* sign
encoding	use of a code to make forms
ethology	the study of animals in their natural habitats
extensional modeling	the extension of primary models both morphologically and connotatively for further representational uses
extensionality principle	principle asserting that higher-order (abstract) models are derivatives of simpler (more concrete) ones
extensionality	process of extending the physical constitution or meaning of forms
externalized form	form made to stand for something

F

fetish	an object that is believed to have magical or spiritual, powers, or which can cause sexual arousal
fiction	text whose content is produced by the imagination and is not necessarily based on fact
firstness	earliest strategy for knowing an object with the senses

focal color	a color category that is universal or associated with a universal sequencing of colors
form	a mental image, or an external representation of something

G

gesticulant	gesture accompanying speech
gesture	use of the hands, the arms, and to a lesser extent, the head, to make bodily forms of all kinds

H

hieroglyphic writing	ancient Egyptian system of writing, in which pictorial symbols were used to represent meaning or sounds or a combination of meaning and sound
holophrase	the one-word utterance produced by infants
hypoicon	icon that is shaped by cultural convention but whose referent can nonetheless be figured out by those who are not members of the culture

I

icon	singularized form which simulates its referent in some way
iconicity	the process of representing referents with iconic forms
image schema	mental impression of locations, movements, shapes, etc.
index	singularized form which establishes a contiguity with its referent (pointing it out, showing its relation to other things, etc.)
indexicality	process of representing referents with indexical signs

indicational (indexical) modeling	representation of referents in terms of their spatiotemporal relation to other referents or to their contexts of occurrence
induction	process of deriving a concept from particular facts of instances
inflection	variations or changes that words undergo to indicate their relations with other words
Innenwelt	the world of internal experiences of a species
intellective code	tertiary code designed to organize knowledge about some field
intercodality	interconnection of one code to other codes
interconnectedness principle	principle claiming that forms are interconnected to each other in various ways
internal model	mental form, mental image
intertextuality	referents present in one text which allude to referents in other texts

L

language	verbal semiosis and representation
layering	the linkage of metaforms to one another

M

map	a representation, usually on a plane surface, of a region of the earth
meaning	particular concept elicited by a specific representational form
mechanical media	media of transmission created technologically (telephones, radios, television sets, computers, videos, etc.)
medium	technical or physical means by which a message is transmitted
meme	replicating pattern of information (tune, idea, clothing fashion, etc.)
mental form	imagined form
mental image	mental outline of something (a shape, a sound, etc.)

meta-metaform	the abstract concept that emerges from the linkage of different metaforms to each other
meta-symbol	symbol derived from the fact that an abstract concept has been associated with source domains that are descriptive of a certain concrete concept
metaform	concept ([thinking = seeing]) that results from the linkage of an abstract notion ([thinking]) with a concrete source domain ([seeing])
metaphor	instantiation of a metaform
metonymy	use of an entity to refer to another that is related to it
mimesis	intentional making of forms in a simulative manner
mode	manner in which a form is encoded (visual, auditory, etc.)
model	form that has been imagined or made externally (through some physical medium) to stand for an object, event, feeling, etc.
modeling	the innate ability to produce forms to represent objects, events, feelings, actions, situations, and ideas perceived to have some meaning, purpose, or useful function
modeling principle	principle claiming that representation is a de facto modeling process
modeling systems theory	theory which posits the presence of species-specific modeling systems that allow a species to produce the forms it needs for understanding the world in its own way
morphological extensionality	process of extending signifiers through some modification of their physical forms
motor program	self-contained circuit able to direct the coordinated movements of many different muscles to accomplish a task
myth	any story or narrative that aims to explain the origin of something
mythology	the study of myths

N

name	form that identifies a human being or, by connotative extension, an animal, an object (such as a commercial product), and event (such as a hurricane)
narrative	something told or written, such as an account, story, tale, etc.
narrator	the teller of the narrative
natural form	form produced by Nature
natural media	natural media of communication such as the voice (speech), the face (expressions), and the body (gesture, posture, etc.)
natural form	a form that is produced by Nature (e.g. a symptom)
novel	a fictional prose narrative of considerable length, typically having a plot that is unfolded by the actions, speech, and thoughts of the characters

O

onomastics	the study of names
onomatopoeia	vocal iconicity (*drip, boom,* etc.)
ontological image schema	image schema underlying the derivation of metaforms perceived to have either the structural properties of [entities], [substances], [containers], [impediments], etc. or the characteristics of physical processes and forms—[plants], [movements], etc.
opposition	process by which forms are differentiated through a minimal change in their signifiers
orientational image schema	image schema underlying the derivation of metaforms perceived to have orientational structure—[up], [down], [back], [front], [near], [far], etc.
osmosis	the spontaneous production of a simulative form in response to some stimulus or need

P

paradigmaticity	a differentiation property of forms
personal deixis	process of referring to the relations that exist among participants taking part in a situation
phoneme	minimal unit of sound in a language that allows its users to differentiate word meanings
phylogenesis	the development of all semiosic abilities (iconicity, symbolism, language, etc.) in the human species
phytosemiosis	semiosis in plants
phytosemiotics	the study of semiosis in plants
pictographic writing	writing in pictures
pictograph	pictorial representation of a referent
pliable model	model that is temporary and adaptive to the dynamics of a situation
primary cohesive modeling	modeling code consisting of particular types of iconic signifiers that serve various simulative representational purposes
primary composite modeling	modeling whereby various iconic signifiers are combined in order to encode complex (non-unitary) referents
primary connective modeling	modeling abstract referents with concrete source domains
primary model	a simulacrum of a referent
primary modeling system	instinctive ability to model the *sensible* properties of things (i.e. properties that can be *sensed*)
primary singularized model	singularized simulative form (icon)
programmed learning	the ability of a species to learn only those things that are relevant to its life needs

R

recognizing stage	the cognitive state whereby a referent is recalled through a form
referent	an object, event, feeling, idea, etc. that is represented by a form

referential domain	a class of objects, events, feelings, ideas, etc. represented by a form
regulator	gesticulant regulating the speech of an interlocutor: e.g. hand movements indicating *Keep going, Slow down,* etc.
representation	process of ascribing a form to some referent
representational principle	principle asserting that knowledge is indistinguishable from how it is represented

S

secondary composite modeling	process of extending a model in either its form or its meaning
secondary connective modeling	process of establishing linkages among already-forged metaforms
secondary model	either an extension of the physical form or meaning of a simulacrum or an indexical form
secondary modeling system	system that allows for indication or the extension of forms
secondary singularized model	model that results either from indication or extension
secondness	ability to refer to the objects through indication or verbal reference
semiosis	capacity of a species to produce and comprehend the specific types of models it requires for processing and codifying perceptual input in its own way
semiotics	the science of signs
sense ratio	level at which one of the senses is activated during the encoding and decoding of forms
Sense-Inference Hypothesis	view that all modeling is initially guided by sensory processes
sensory modeling	creating mental models through the senses
sign	something that stands for something else (singularized form)
sign stimulus (releaser)	cue that enables animals to recognize a critical referent when they encounter it for the first time

signal	sign that naturally or conventionally (artificially) triggers some reaction on the part of a receiver
signification	relation that holds between a form and its referent
signified	part of a sign that is referred to (the referent)
signifier	part of a sign that does the referring (the form)
signifying order	interconnected system of signs, texts, codes, and connective forms
simulacrum	simulated form
singularized form	form that has been created to stand for a singular referent or referential domain
social code	tertiary code intended to regulate human interaction
sociobiology	the study of biological evolution in terms of its codependency with social and cultural evolution in all species
source domain	set of vehicles (concrete forms) that is used to deliver the meaning of an abstract concept
spatial deixis	process of referring to the spatial locations of referents
speech	expressed language
stable model	model that is fixed and relatively permanent or invariable
structural image schema	image schema that combines ontological and orientational properties
structuralism	the approach in semiotics that views signs as reflexes of intellectual and emotional structures in the human psyche
structuralist principle	principle asserting that all forms display the same pattern of structural properties
structure	any repeatable or predictable aspect of models
subordinate concept	concept needed for specialized purposes
superordinate concept	concept with a highly general referential function
symbol	singularized form that stands arbitrarily or conventionally for its referent

symbolicity	the process of representing referents with symbolic forms
symbolism	symbolic meaning in general
symptom	natural sign designed to alert an organism to the presence of altered states in its body
synchronicity	refers to the fact that forms are constructed at a given point in time for some particular purpose or function
syndrome	configuration of symptoms with a fixed denotatum
syntagmaticity	combinatory property of forms
syntax	syntagmatic structure in language
systems analysis	the study of modeling systems across species

T

target domain	what a metaform is about (abstract concept that is metaphorized)
technology	processes by which human beings fashion objects in order to increase their understanding of, and control over, the material environment
temporal deixis	process of referring to the temporal relations that exist among things and events
tertiary cohesive modeling	modeling of referents with symbolic forms
tertiary composite modeling	modeling composite referents with symbolic forms
tertiary connective modeling	linkages among metaforms and concrete concepts from which meta-symbols can be extracted
tertiary model	a symbolically-devised form
tertiary modeling system	modeling system that undergirds highly abstract, symbol-based modeling
tertiary singularized model	single symbol
text	something put together to represent complex (non-unitary) referents
thirdness	abstract form of knowing
transmission	the sending and reception of messages

U

Umwelt	domain that a species is capable of modeling (the external world of experience to which a species has access)

Z

zoosemiosis	semiosis in animals
zoosemiotics	the study of semiosis in animals

Works cited and general bibliography

Antenos-Conforti, Enza, Edward Barbeau, and Marcel Danesi

1997 *Problem-Solving in Mathematics: A Semiotic Perspective with Educational Implications*. Toronto, Welland, and Lewsiton: Soleil Publishers (Monograph Series of the Toronto Semiotic Circle, volume 20).

Appelbaum, David

1990 *Voice*. Albany: State University of New York Press.

Ardrey, Robert

1966 *The Territorial Imperative*. New York: Atheneum.

Argyle, Michael

1988 *Bodily Communication*. New York: Methuen.

Aristotle

1952 *Poetics*. In: *The Works of Aristotle*, vol. 11, W. D. Ross (ed.). Oxford: Clarendon Press.

Armstrong, R. L.

1965 John Locke's doctrine of signs: A new metaphysics. *Journal of the History of Ideas* 26: 369-382.

Arnheim, Rudolf

1969 *Visual Thinking*. Berkeley: University of California Press.

Asch, Solomon

1950 On the use of metaphor in the description of persons. In: H. Werner (ed.), *On Expressive Language*, 86–94. Worcester: Clark University Press.

1958 The metaphor: A psychological inquiry. In: R. Tagiuri and L. Petrullo (eds.), *Person Perception and Interpersonal Behavior*, 28-42. Stanford: Stanford University Press.

Axtell, Roger E.

1991 *Gestures*. New York: John Wiley.

Ayer, A. J.

1968 *The Origins of Pragmatism: Studies in the Philosophy of Charles Sanders Peirce and William James*. London: Macmillan.

Baer, E.

1982 The medical symptom: Phylogeny and ontogeny. *American Journal of Semiotics* 1: 17-34.

Baigrie, Brian S. (ed.)

1996 *Picturing Knowledge: Historical and Philosophical Problems Concerning the Use of Art in Science.* Toronto: University of Toronto Press.

Bal, Mieke

1985 *Narratology: Introduction to the Theory of the Narrative.* Toronto: University of Toronto Press.

Bally, Charles

1939 Qu'est-ce qu'un signe? *Journal de Psychologie Normale et Pathologique* 112: 161-174.

Bar-Hillel, Y.

1954 Indexical expressions. *Mind* 63: 359-379.

1970 *Aspects of Language*. Jerusalem: The Magnes Press.

Barlow, Horace, Colin Blakemore, and Miranda Weston-Smith (eds.)

1990 *Images and Understanding.* Cambridge: Cambridge University Press.

Barthes, Roland

1957 *Mythologies.* Paris: Seuil.

1964 *Éléments de sémiologie.* Paris: Seuil.

1972 Sémiologie et médicine. In: R. Bastide (ed.), *Les sciences de folie*, 37-46. Paris: Mouton.

Barwise, J. and J. Perry

1983 *Situations and Attitudes*. Cambridge, Mass.: MIT Press.

Bateson, Gregory

1968 *Animal Communication: Techniques of Study and Results of Research*. Bloomington: Indiana University Press.

Beaudrillard, Jean

1981 *For a Critique of the Political Economy of the Sign.* St. Louis: Telos Press.

Becker, N. and E. Schorsch

1975 Geldfetischismus. In: E. Schorsch and G. Schmidt (eds.), *Ergebnisse zur Sexualforschung*, 238-256. Cologne: Wissenschafts-Verlag.

Behan, R. J.

1926 *Pain: Its Origin, Conduction, Perception, and Diagnostic Significance*. New York: Appleton.

Bemporad, J., D. Dunton, and F. H. Spady

1976 Treatment of a child foot fetishist. *American Journal of Psychotherapy* 30: 303-316.

Benveniste, Emile

1971 *Problems in General Linguistics*. Coral Gables: University of Miami Press.

Berg, H. C.

1976 Does the flagellar rotary motor step? *Cell Motility* 3: 47-56.

Berger, John

1972 *Ways of Seeing*. Harmondsworth: Penguin.

Bergin, Thomas G. and Max H. Fisch

1984 *The New Science of Giambattista Vico*. Ithaca: Cornell University Press.

Berlin, Brent and Paul Kay

1969 *Basic Color Terms*. Berkeley: University of California Press.

Bernardelli, Andrea (ed.)

1997 *The Concept of Intertextuality Thirty Years On: 1967-1997*. Special Issue of *Versus*, 77/78. Milano: Bompiani.

Bickerton, Derek

1981 *The Roots of Language*. Ann Arbor: Karoma Publishers.

1990 *Language and Species*. Chicago: University of Chicago Press.

Bilz, R.

1940 *Pars pro toto*. Leipzig: Georg Thieme.

Birdwhistell, R.

1970 *Kinesics and Context: Essays on Body Motion Communication*. Harmondsworth: Penguin.

Black, Max

1962 *Models and Metaphors*. Ithaca: Cornell University Press.

Bloomfield, Leonard

1933 *Language*. New York: Holt.

1939 Linguistic aspects of science. *International Encyclopedia of Unified Science* 1: 215-278.

Bonner, John T.

1980 *The Evolution of Culture in Animals*. Princeton: Princeton University Press.

Bornet, J.

1892 *Early Greek Philosophy*. London: MacMillan.

Bouissac, Paul et al. (eds.)

1986 *Iconicity: Essays on the Nature of Culture*. Tübingen: Stauffenberg.

Boysson-Bardies, B. de and M. M. Vihman

1991 Adaptation to language: Evidence from babbling and first words in four languages. *Language* 67: 297-319.

Braten, S.

1988 Dialogic mind: The infant and the adult in protoconversation. In: M. E. Carvallo (ed.), *Nature, Cognition, and System I*, 187-205. Dordrecht: Kluwer.

Bremer, J. and H. Roodenburg (eds.)

1991 *A Cultural History of Gesture.* Ithaca: Cornell University Press.

Britton, John

1970 *Language and Learning.* Harmondsworth: Penguin.

Bronowski, Jacob

1967 Human and animal language. In: *To Honor Roman Jakobson*, 374-394. The Hague: Mouton.

Brosses, C. de

1760 *La culte des dieux fétiches*. Paris.

Brown, Roger W.

1958a How shall a thing be called? *Psychological Review* 65: 14-21.

1958b *Words and Things*. New York: Free Press.

1970 *Psycholinguistics*. New York: Free Press.

1986 *Social Psychology*. New York: Free Press.

Brown, R. W., R. A. Leiter, and D. C. Hildum

1957 Metaphors from music criticism. *Journal of Abnormal and Social Psychology* 54: 347-352.

Bruner, Jerome

1983 *Children's Talk: Learning to Use Language*. New York: W. W. Norton.

Bühler, Karl

1908 On thought connection. In: D. Rapaport (ed.), *Organization and Pathology of Thought*, 81-92. New York: Columbia University Press.

1934 *Sprachtheorie: Die Darstellungsfunktion der Sprache*. Jena: Fischer.

Bullowa, M. (ed.)

1979 *Before Speech: The Beginning of Interpersonal Communication.* Cambridge: Cambridge University Press.

Burkhardt, D. et al.

1967 *Signals in the Animal World.* New York: McGraw-Hill.

Bursill-Hall, G. L.

1963 Some remarks on deixis. *Canadian Journal of Linguistics* 8: 82-96.

Butler, C.

1970 Chemical communication in insects: Behavioral and ecological aspects. *Communication by Chemical Signals* 1: 35-78.

Buyssens, E.

1943 *Les langages et le discours*. Brussels: Office de Publicité.

Carnap, Rudolf

1942 *Introduction to Semantics*. Cambridge, Mass.: Harvard University Press.

1956 *Meaning and Necessity: A Study in Semantics and Modal Logic*. Chicago: University of Chicago Press.

Carpenter, C. R.

1969 Approaches to studies of the naturalistic communicative behavior in nonhuman primates. In: T. A. Sebeok et al. (eds.), *Approaches to Animal Communication*, 40-70. The Hague: Mouton.

Cassirer, Ernst

1944 *An Essay on Man: An Introduction to a Philosophy of Human Culture*. New Haven: Yale University Press.

1946 *Language and Myth*. New York: Dover.

1957 *The Philosophy of Symbolic Forms*. New Haven: Yale University Press.

Celon, E. and S. Marcus

1973 Le diagnostic comme langage. *Cahiers de Linguistique* 10: 163-173.

Chadwick, J. and W. N. Mann

1950 *The Medical Works of Hippocrates*. Oxford: Blackwell.

Chamberlain, E. N. and C. Ogilvie

1974 *Symptoms and Signs in Clinical Medicine*. Bristol: Wright.

Chao, Y. R.

1962 Models in linguistics and models in general. In: E. Nagel, P. Suppes, and A. Tarski (eds.), *Logic, Methodology, and the Philosophy of Science*, 558-566. Stanford: Stanford University Press.

1968 *Language and Symbolic Systems*. Cambridge: Cambridge University Press.

Cheraskin, E. and W. Rinsdorf

1973 *Predictive Medicine: A Study in Strategy*. Mountainview, Ca.: Pacific Press.

Cherry, Colin

1966 *On Human Communication*. Cambridge, Mass.: MIT Press.

Cherwitz, Richard and James Hikins

1986 *Communication and Knowledge: An Investigation in Rhetorical Epistemology*. Columbia: University of South Carolina Press.

Chomsky, Noam

1957 *Syntactic Structures*. The Hague: Mouton.

1976 On the nature of language. In: H. B. Steklis, S. R. Harnad, and J. Lancaster (eds.), *Origins and Evolution of Language and Speech*, 46-57. New York: New York Academy of Sciences.

1980 *Rules and Representations*. New York: Columbia University Press.

1986 *Knowledge of Language: Its Nature, Origin, and Use*. New York: Praeger.

1990 Language and mind. In: D. H. Mellor (ed.), *Ways of Communicating*, 56-80. Cambridge: Cambridge University Press.

Clarke, D. S.

1987 *Principles of Semiotic*. London: Routledge and Kegan.

Colby, K. M. and M. T. McGuire

1981 Signs and symptoms. *The Sciences* 21: 21-23.

Colton, Helen

1983 *The Gift of Touch*. New York: Putnam.

Copeland, J. E. (ed.)

1984 *New Directions in Linguistics and Semiotics*. Houston: Rice University Studies.

Coseriu, E.

1967 L'arbitraire du signe: zur Spätgeschichte eines aristotelischen Begriffes. *Archiv für das Studium der Neueren Sprachen und Literaturen* 204: 81-112.

Count, E. W.

1969 Animal communication in man-science. In: Thomas A. Sebeok et al. (eds.), *Approaches to Animal Communication*, 71-130. The Hague: Mouton.

Cox, Maureen

1992 *Children's Drawings*. Harmondsworth: Penguin.

Craig, Colette (ed.)

1986 *Noun Classes and Categorization*. Amsterdam: John Benjamins.

Croft, W.

1991 *Syntactic Categories and Grammatical Relations*. Chicago: University of Chicago Press.

Crookshank, F. G.

1925 The importance of a theory of signs and a critique of language in the study of medicine. In: *The Meaning of Meaning*, supplement II. London: Kegan Paul.

Crystal, David

1987 *The Cambridge Encyclopedia of Language*. Cambridge: Cambridge University Press.

Culler, Jonathan

1983 *Roland Barthes*. New York: Oxford University Press.

Danesi, Marcel

1990 Thinking is seeing: Visual metaphors and the nature of abstract thought. *Semiotica* 80: 221-237.

1993 *Vico, Metaphor, and the Origin of Language*. Bloomington: Indiana University Press.

1995 *Giambattista Vico and the Cognitive Science Enterprise*. New York: Peter Lang.

1998 The dimensionality principle and semiotic analysis. *Sign Systems Studies* 26: 42-60.

1999a The dimensionality of metaphor. *Sign Systems Studies* 27: 54-69.

1999b The interconnectedness principle and the analysis of discourse. *Applied Semiotics* 6/7: 393-399.

Danesi, Marcel and Donato Santeramo

1995 *Deictic Verbal Constructions*. Urbino: Centro Internazionale di Semiotica e di Linguistica.

Dante Alighieri

1957 *De vulgari eloquentiae*, A. Marigo (ed.). Firenze: Le Monnier.

Darwin, Charles

1859 *The Origin of Species*. New York: Collier.

1871 *The Descent of Man*. New York: Modern Library.

1872 *The Expression of the Emotions in Man and Animals*. London: John Murray.

Davis, Paul J. and Reuben Hersh

1986 *Descartes' Dream: The World according to Mathematics*. Boston: Houghton Mifflin.

Dawkins, Richard

1976 *The Selfish Gene*. Oxford: Oxford University Press.

1987 *The Blind Watchmaker*. Harlow: Longmans.

1995 *River Out of Eden: A Darwinian View of Life*. New York: Basic.

De Lacy, P. H. and E. A. De Lacy

1941 *Philodemeus on Methods of Inference*. Philadelphia: American Philological Association.

De Laguna, G. A.

1927 *Speech: Its Function and Development*. Bloomington: Indiana University Press.

Deacon, Terrence W.

1997 *The Symbolic Species: The Co-Evolution of Language and the Brain*. New York: Norton.

Deane, Paul

1992 *Grammar in Mind and Brain: Explorations in Cognitive Syntax*. Berlin: Mouton de Gruyter.

Deely, John

1980 *The Signifying Animal: The Grammar of Language and Experience*. Bloomington: Indiana University Press.

1982 *Introducing Semiotics*. Bloomington: Indiana University Press.

1985 Semiotic and the liberal arts. *The New Scholasticism* 59: 296-322.

1990 *Basics of Semiotics*. Bloomington: Indiana University Press.

Dennett, Daniel C.

1991 *Consciousness Explained*. Boston: Little, Brown.

Descartes, René

1637 *Essaies philosophiques*. Leyden: L'imprimerie de Ian Maire.

Douglas, Mary

1992 *Objects and Objections*. Toronto: Monographs of the Toronto Semiotic Circle.

Dubois, P.

1988 *L'acte photographique*. Brussels: Labor.

Dundes, Alan

1972 Seeing is believing. *Natural History* 81: 9-12.

Dunning, W. V.

1991 *Changing Images of Pictorial Space: A History of Visual Illusion in Painting*. Syracuse: Syracuse University Press.

Eccles, John, C.

1979 *The Human Mystery*. New York: Springer.

1992 *The Human Psyche*. London: Routledge.

Eco, Umberto

1972a *Einführung in die Semiotik*. München: Fink.

1972b Introduction to a semiotics of iconic signs. *VS: Quaderni di Studi Semiotici* 2: 1-15.

1976 *A Theory of Semiotics*. Bloomington: Indiana University Press.

1977 The influence of Roman Jakobson on the development of semiotics. In: D. Armstrong and C. H. van Schoonefeld (eds.), *Roman Jakobson: Echoes of His Scholarship*, 39-58. Lisse: The Peter de Ridder Press.

1980 The sign revisited. *Philosophy and Social Criticism* 7: 261-297.

1984 *Semiotics and the Philosophy of Language*. Bloomington: Indiana University Press.

Eco, Umberto and Thomas A. Sebeok (eds.)

1983 *The Sign of Three: Dupin, Holmes, Peirce*. Bloomington: Indiana University Press.

Efron, D.

1972 *Gesture, Race, and Culture*. The Hague: Mouton.

Ekman, Paul

1985 *Telling Lies*. New York: Norton.

Ekman, Paul and Walter Friesen

1969 The repertoire of nonverbal behavior: Categories, origins, usage, and coding. *Semiotica* 1: 49-98.

1975 *Unmasking the Face*. Englewood Cliffs: Prentice-Hall.

Elstein, A. S. et al.

1978 *Medical Problem Solving: An Analysis of Clinical Reasoning*. Cambridge, Mass.: Harvard University Press.

Emerson, A. E.

1938 Termite nests: A study of the phylogeny of behavior. *Ecological Monographs* 8: 247-284.

Engen, T.

1982 *The Perception of Odours*. New York: Academic.

Engler, R.

1962 Théorie et critique d'un principe saussurien: l'arbitraire du signe. *CFS* 19: 5-66.

Enninger, Werner

1992 Clothing. In: R. Bauman (ed.), *Folklore, Cultural Performances, and Popular Entertainments*, 123-145. Oxford: Oxford University Press.

Ennion, E. R. and Nikolaas Tinbergen

1967 *Tracks*. Oxford: Oxford University Press.

Erckenbrecht, U.

1976 *Das Geheimnis des Fetischismus. Grundmotive der Marxschen Erkenntniskritik.* Frankfurt am Main: Europäische Verlagsanstalt.

Fabrega, H.

1974 *Disease and Social Behavior: An Interdisciplinary Perspective.* Cambridge, Mass.: MIT Press.

Fauconnier, Gilles

1985 *Mental Spaces*. Cambridge: Cambridge University Press.

1997 *Mappings in Thought and Language*. Cambridge: Cambridge University Press.

Fauconnier, Gilles and Eve Sweetser (eds.)

1996 *Spaces, Worlds, and Grammar*. Chicago: University of Chicago Press.

Feher, M., R. Naddaf, and N. Tazi (eds.)

1989 *Fragments for a History of the Human Body*. New York: Zone.

Fillmore, Charles J.

1972 A grammarian Looks at sociolinguistics. *Georgetown University Monograph Series in Languages and Linguistics* 25: 273-287.

1973 May We Come In? *Semiotica* 9: 97-116.

1997 *Lectures on Deixis*. Stanford: CSLI Publications.

Fisch, Max H.

1978 Peirce's general theory of signs. In: Thomas A. Sebeok (ed.), *Sight, Sound, and Sense*, 31-70. Bloomington: Indiana University Press.

1980 Foreword. In: Thomas A. Sebeok and Jean Umiker-Sebeok (eds.), *You Know My Method*, 7-13. Bloomington: Gaslight Publications.

Fisher, Helen E.

1992 *Anatomy of Love*. New York: Norton.

Fiske, John C.

1979 *Introduction to Communication Studies*. London: Methuen.

Foucault, Michel

1976 *The History of Sexuality*, vol. 1. London: Allen Lane.

Fox, J. J.

1975 On binary categories and primary symbols: Some Rotinese perspectives. In: R. Willis (ed.), *The Interpretation of Symbolism*, 99-132. New York: John Wiley and Sons.

Freedman, A. M., H. I. Kaplan, and B. J. Sadock

1972 *Modern Synopsis of Comprehensive Textbook of Psychiatry*. Baltimore: Williams & Wilkins.

Frege, Gottlob

1892 Über Sinn und Bedeutung. *Zeitschrift für Philologie und philologische Kritik* 100: 25-50.

Frei, H.

1944 Systèmes de déictiques. *AI* 4: 111-129.

1950 Zéro, vide et intermittent. *Zeitschrift für Phonologie* 4: 161-191.

French, A. P. and P. J. Kennedy (eds.)

1985 *Niels Bohr: A Centenary Volume*. Cambridge, Mass.: Harvard University Press.

Freud, Sigmund

1927 Fetishism. In: *The Standard of Edition of the Complete Psychological Works* 21: 149-157.

Friedmann, H.

1955 The honey-guides. *U.S. National Museum Bulletin* 208. Washington, D.C.: Smithsonian.

Frisch, Karl von

1962 Dialects in the language of bees. *Scientific American* 207: 79-87.

1967 *The Dance Language and Orientation of Bees*. Cambridge, Mass.: Harvard University Press.

Frisch, Karl von and Otto von Frisch

1974 *Animal Architecture*. New York: Harcourt.

Frutiger, A.

1989 *Signs and Symbols*. New York: Van Nostrand.

Frye, Northrop

1981 *The Great Code: The Bible and Literature*. Toronto and New York: Academic Press.

Furnham, A.

1988 Write and wrong: The validity of graphological analysis. *The Skeptical Inquirer* 13: 64-69.

Gale, Richard M.

1967 Indexical signs, egocentric particulars, and token-reflexive words. *The Encyclopedia of Philosophy* 4: 151-155.

Gardiner, A. H.

1932 *The Theory of Speech and Language*. Oxford: Clarendon Press.

Gardner, B. T. and R. A. Gardner

1975 Evidence for sentence constituents in the early utterances of child and chimpanzee. *Journal of Experimental Psychology* 104: 244-262.

Gardner, Martin

1968 *Logic Machines, Diagrams, and Boolean Algebra*. New York: Dover.

Gardner, R. A. and B. T. Gardner

1969 Teaching sign language to a chimpanzee. *Science* 165: 664-672.

Garnier, P.

1896 *Fétischistes: Pervertis et invertis sexuels.* Paris.

Garver, N.

1986 Review of Shapiro 1983. *Transactions of the Charles S. Peirce Society* 22: 68-74.

Gebhard, P. H.

1969 Fetishism and sadomaschosim. *Science and Psychoanalysis* 15: 71-80.

Genette, G.

1988 *Narrative Discourse Revisited.* Ithaca: Cornell University Press.

Geras, N.

1971 Essence and appearance: aspects of fetishism in Marx's *Capital. New Left Review* 65: 69-85.

Gessinger, J. and W. von Rahden (eds.)

1988 *Theorien vom Ursprung der Sprache.* Berlin: Walter de Gruyter.

Gibbs, Raymond W.

1994 *The Poetics of Mind: Figurative Thought, Language, and Understanding*. Cambridge: Cambridge University Press.

Ginzburg, C.

1983 Morelli, Freud and Sherlock Holmes. In: Umberto Eco and Thomas A. Sebeok (eds.), *The Sign of Three,* 81-118. Bloomington: Indiana University Press.

Gipper, H.

1963 *Bausteine zur Sprachinhaltforschung: Neuere Sprachbetrachtung im Austausch mit Geistes- und Naturwissenschaft.* Düsseldorf: Pädagogischer Verlag Schwann.

Goatley, Andrew

1997 *The Language of Metaphors.* London: Routledge.

Godel, R.

1953 La question des signes zéro. *Cahiers Ferdinand de Saussure* 11: 31-41.

Goffman, Erving

1959 *The Presentation of Self in Everyday Life.* New York: Anchor.

1963 *Stigma: Notes on the Management of Spoiled Identity.* Englewood Cliffs: Prentice-Hall.

Gombrich, E. H.

1951 Meditations on a hobby horse or the roots of artistic form. In: L. L. Whyte (ed.), *Aspects of Form*, 209-228. Bloomington: Indiana University Press.

Goode, J.

1992 Food. In: R. Bauman (ed.), *Folklore, Cultural Performances, and Popular Entertainments*, 233-245. Oxford: Oxford University Press.

Goody, J.

1982 *Cooking, Cuisine and Class.* Cambridge: Cambridge University Press.

Goossens, L. et al.

1995 *By Word of Mouth: Metaphor, Metonymy and Linguistic Action in a Cognitive Perspective.* Berlin: Mouton de Gruyter.

Gould, Stephen J. and E. S. Vrba

1982 Exaptation: A missing term in the science of form. *Paleobiology* 8: 4-15.

Greenberg, Joseph H.

1987 *Language in the Americas.* Stanford: Stanford University Press.

Greenbie, B.

1981 *Spaces: Dimensions of the Human Landscape.* New Haven: Yale University Press.

Greimas, Algirdas J.

1987 *On Meaning: Selected Essays in Semiotic Theory*, trans. by P. Perron and F. Collins. Minneapolis: University of Minnesota Press.

Greimas, Algirdas J. and J. Courtés

1979 *Semiotics and Language*. Bloomington: Indiana University Press.

Griffin, Donald R.

1981 *The Question of Animal Consciousness*. New York: Rockefeller University Press.

1992 *Animal Minds*. Chicago: University of Chicago Press.

Guthrie, R. Dale

1976 *Body Hot Spots: The Anatomy of Human Social Organs and Behavior*. New York: Van Nostrand Reinhold.

Haas, W.

1957 Zero in linguistic description. *Studies in Linguistic Analysis*, Philological Society of London, 35-53. Oxford: Blackwell.

Hadamard, J.

1945 *An Essay on the Psychology of Invention in the Mathematical Field.* Princeton: Princeton University Press.

Haldane, J. B. S.

1955 Animal communication and the origin of human language. *Science Progress* 43: 385-401.

Hall, Edward T.

1966 *The Hidden Dimension.* New York: Doubleday.

1973 *The Silent Language*. New York: Anchor.

Hall, K. R. L. and De Vore, I.

1965 Baboon social behavior. In: I. Devore (ed.), *Primate Behavior*, 53-110. New York: Holt, Rinehart and Winston.

Halliday, M. A. K.

1975 *Learning How to Mean: Explorations in the Development of Language*. London: Arnold.

1985 *Introduction to Functional Grammar*. London: Arnold.

Haraway, D.

1989 *Primate Visions: Gender, Race and Nature in the World of Modern Science.* London: Routledge.

Hardin, C. L. and L. Maffi (eds.).

1997 *Color Categories in Thought and Language.* Cambridge: Cambridge University Press.

Harnad, S. R., H. B. Steklis, and J. Lancaster (eds.)

1976 *Origins and Evolution of Language and Speech.* New York: New York Academy of Sciences.

Harré, Rom

1981 *Great Scientific Experiments*. Oxford: Phaidon Press.

Hausman, Carl R.

1989 *Metaphor and Art*. Cambridge: Cambridge University Press.

Hawkes, T.

1977 *Structuralism and Semiotics*. Berkeley: University of California Press.

Hearne, V.

1986 *Adam's Task: Calling Animals by Name*. New York: Knopf.

Hediger, H.

1967 Verstehens- und Verständigungsmöglichkeiten zwischen Mensch und Tier. *Schweizerische Zeitschrift für Psychologie und ihre Anwendungen* 26: 234-255.

1968 *The Psychology and Behaviour of Animals in Zoos and Circuses*. New York: Dover.

Heidel, W. A.

1941 *Hippocratic Medicine: Its Spirit and Method*. New York: Columbia University Press.

Heine, B.

1997 *Cognitive Foundations of Grammar*. Oxford: University Press.

Heisenberg, Werner

1949 *The Physical Principles of the Quantum Theory*. New York: Dover.

Herskovits, M.

1948 *Man and His Works*. New York: Alfred A. Knopf.

Hesse, M.

1967 Models and analogy in science. *The Encyclopedia of Philosophy* 5: 354-359.

Hewes, G. W.

1973 Primate communication and the gestural origin of language. *Current Anthropology* 14: 5-24.

1974 *Language Origins: A Bibliography*. The Hague: Mouton.

Hinton, H. E.

1973 Natural deception. In: R. L. Gregory and E. H. Gombrich (eds.), *Illusion in Nature and Art*, 97-159. London: Duckworth.

Hinton, L., J. Nichols, and J. J. Ohala (eds.)

1994 *Sound Symbolism*. Cambridge: Cambridge University Press.

Hjelmslev, Louis

1963 *Prolegomena to a Theory of Language*. Madison: University of Wisconsin Press.

Hobbes, Thomas

1656 *Elements of Philosophy*. London: Molesworth.

Hockett, Charles F.

1960 The origin of speech. *Scientific American* 203: 88-96.

Hoffmeyer, Jesper

1996 *Signs of Meaning in the Universe*. Bloomington: Indiana University Press.

Hollander, Anne

1978 *Seeing through Clothes*. Harmondsworth: Penguin.

Hollander, J.

1959 The metrical emblem. *Kenyon Review* 21: 279-296.

Honeck, R. P. and Robert R. Hoffman (eds.)

1980 *Cognition and Figurative Language*. Hillsdale, N. J.: Lawrence Erlbaum.

Howes, David

1999 The facts of sense: A semiotics of perception and sensory anthropology. *International Journal of Applied Semiotics* 1: 141-158.

Hudson, Liam

1972 *The Cult of the Fact*. New York: Harper & Row.

Humboldt, Wilhelm von

1836 [1988] *On Language: The Diversity of Human Language-Structure and Its Influence on the Mental Development of Mankind*, P. Heath (trans. Cambridge: Cambridge University Press.

Humphries, W. C.

1968 *Anomalies and Scientific Theories*. San Francisco: Freeman.

Husserl, Edmund

1970 *Philosophie der Arithmetik*, L. Eley (ed.). The Hague: Nijhoff.

Huxley, Julian

1966 A discussion of ritualization of behaviour in animals and men. *Philosophical Transactions of the Royal Society of London* 251: 247-526.

Hymes, Dell

1971 *On Communicative Competence*. Philadelphia: University of Pennsylvania Press.

Indurkhya, B.

1992 *Metaphor and Cognition*. Dordrecht: Kluwer.

Ingram, D.

1978 Typology and universals of personal pronouns. In: J. H. Greenberg (ed.), *Universals of Human Language*, 213-247. Stanford: Stanford University Press.

Inhelder, B. and Jean Piaget

1958 *The Growth of Logical Thinking from Childhood through Adolescence*. New York: Basic.

Jackendoff, Ray

1994 *Patterns in Mind: Language and Human Nature*. New York: Basic.

Jacob, F.

1974 *The Logic of Living Systems: A History of Heredity*. London: Allen Lane.

1982 *The Possible and the Actual*. Seattle: University of Washington Press.

Jakobson, Roman

1960 Linguistics and poetics. In: Thomas A. Sebeok (ed.), *Style and Language*, 34-45. Cambridge, Mass.: MIT Press.

1963 *Essais de linguistique générale*. Paris: Editions de Minuit.

1965 Quest for the essence of language. *Diogenes* 51: 21-37.

1966 Signe zero. In: E. Hamp et al. (eds.), *Readings in Linguistics II*, 109-115. Chicago: University of Chicago Press.

1970 Language in relation to other communication systems. In: C. Olivetti (ed.), *Linguaggi nella società e nella tecnica*, 3-16. Milano: Edizioni di Comunità.

1971 *Selected Writings II: Word and Language*. The Hague: Mouton.

1974 *Main Trends in the Science of Language*. New York: Harper & Row.

1980 *The Framework of Language*. Ann Arbor: Michigan Studies in the Humanities.

Jarvella R. H. and W. Klein (eds.)

1982 *Speech, Place and Action: Studies in Deictic and Related Topics*. New York: John Wiley and Sons.

Jastrow, J.

1930 *A History of Psychology in Autobiography* 1: 135-162.

Jaynes, Julian

1976 *The Origin of Consciousness in the Breakdown of the Bicameral Mind*. Toronto: University of Toronto Press.

Jerne, N. K.
1985 The generative grammar of the immune system. *Science* 229: 1057-9.

Jespersen, Otto
1964 *Language, Its Nature, Development, and Origin*. New York: Norton.

Johnson, Mark
1987 *The Body in the Mind: The Bodily Basis of Meaning, Imagination and Reason*. Chicago: University of Chicago Press.

Johnson-Laird, P. N.
1983 *Mental Models*. Cambridge, Mass.: Harvard University Press.

Jones, R.
1982 *Physics as Metaphor*. New York: New American Library.

Jung, Carl
1956 *Analytical Psychology*. New York: Meridian.

Kahn, T. C.
1969 Symbols and man's nature. *The International Journal of Symbolology* 1: 5-6.

Kant, Immanuel
1790 *Critique of Judgment*. New York: Hafner Press.

Kantor, J. R.
1936 *An Objective Psychology of Grammar*. Bloomington: Indiana University Press.

Kecskemeti, P.
1952 *Meaning, Communication, and Value*. Chicago: University of Chicago Press.

Kendon, A.
1991 Some considerations for a theory of language origins. *Man* 26: 199-221.

Kevles, D. J.
1985 *In the Name of Eugenics: Genetics and the Uses of Human Heredity*. New York: Knopf.

Kinsey, A. C., W. B. Pomeroy, C. E. Marshall, and P. H. Gebhard
1953 *Sexual Behavior in the Human Female*. Philadelphia: Saunders.

Kinzle, D.
1982 *Fashion and Fetishism: A Social History of the Corset, Tight-Lacing and Other Forms of Body-Sculpture in the West*. Totowa: Rowman and Littlefield.

Kleinpaul, R.

1972 [1888] *Sprache ohne Worte: Idee einer allgemeinen Wissenschaft der Sprache*. The Hague: Mouton.

Kloft, W.

1959 Versuch einer Analyse der Trophobiotischen Beziehungen von Ameisen zu Aphiden, *Biologisches Zentralblatt* 78: 863-870.

Koch, W. A.

1986 *Philosophie der Philologie und Semiotik*. Bochum: Brockmeyer.

Koch, W. A. (ed.)

1989 *Geneses of Language*. Bochum: Brockmeyer.

Köhler, Wolfgang

1925 *The Mentality of Apes*. London: Routledge and Kegan Paul.

Konner, Melvin

1987 On human nature: Love among the robots. *The Sciences* 27: 14-23.

1991 Human nature and culture: Biology and the residue of uniqueness. In: J. J. Sheehan and M. Sosna (eds.), *The Boundaries of Humanity*, 103-124. Berkeley: University of California Press.

Kosslyn, Stephen M.

1983 *Ghosts in the Mind's Machine: Creating and Using Images in the Brain*. New York: W. W. Norton.

Kövecses, Z.

1986 *Metaphors of Anger, Pride, and Love: A Lexical Approach to the Structure of Concepts*. Amsterdam: Benjamins.

1988 *The Language of Love: The Semantics of Passion in Conversational English*. London: Associated University Presses.

1990 *Emotion Concepts*. New York: Springer.

Krafft-Ebing, Richard von

1886 *Psychopathia sexualis*. Stuttgart.

Krampen, Martin

1981 Phytosemiotics. *Semiotica* 36: 187-209.

1991 *Children's Drawings*. New York: Plenum.

Kuhn, C. G.

1821-33 *Claudii Galeni opera omnia*. Leipzig: Cnobloch.

Kuhn, Thomas S.

1970 *The Structure of Scientific Revolutions*. Chicago: University of Chicago Press.

Labov, William and D. Fanshel

1977 *Therapeutic Discourse.* New York: Academic.

Laitman, J. T

1983 The evolution of the hominid upper respiratory system and implications for the origins of speech. In: E. de Grolier (ed.), *Glossogenetics: The Origin and Evolution of Language*, 63-90. Utrecht: Harwood.

1990 Tracing the origins of human speech. In: P. Whitten and D. E. K. Hunter (eds.), *Anthropology: Contemporary Perspectives*, 124-130. Glenview, Ill.: Scott, Foresman and Company.

Lakoff, George

1987 *Women, Fire and Dangerous Things: What Categories Reveal about the Mind.* Chicago: University of Chicago Press.

Lakoff, George and Mark Johnson

1980 *Metaphors We Live By.* Chicago: Chicago University Press.

1999 *Philosophy in the Flesh: The Embodied Mind and Its Challenge to Western Thought.* New York: Basic.

Lakoff, George and Mark Turner

1989 *More than Cool Reason: A Field Guide to Poetic Metaphor.* Chicago: University of Chicago Press.

Lambert, J. H.

1764 *Semiotik oder Lehre von der Bezeichnung der Gedanken und Dinge.* Leipzig: Johann Wendler.

Landar, H.

1966 *Language and Culture.* Oxford: Oxford University Press.

Landsberg, Marge E. (ed.)

1988 *The Genesis of Language: A Different Judgement of Evidence.* Berlin: Mouton de Gruyter.

Langacker, Ronald W.

1987 *Foundations of Cognitive Grammar.* Stanford: Stanford University Press.

1990 *Concept, Image, and Symbol.* Berlin: Mouton de Gruyter.

Langer, Susanne K.

1948 *Philosophy in a New Key.* Cambridge: Harvard University Press.

1957 *Problems of Art.* New York: Scribner's.

Larker, M. (ed.)

1968 *Bibliographie zur Symbolik, Ikonographie und Mythologie.* Baden-Baden: Heitz.

Latham, R. G.

1848 *The Works of Thomas Sydenham, M.D.* London: Sydenham Society.

Lausberg, Heinrich

1960 *Handbuch der Literarischen Theorie.* Munich: Max Hueber.

Lawick-Goodall, Jane

1968 The behaviour of free-living chimpanzees in the Gombe Stream Reserve. *Animal Behaviour Monographs* 1: Part 3.

Lawrence, C.

1982 Illnesses and their meanings. *Times Literary Supplement,* p. 148.

Leach, Edmund

1976 *Culture and Communication.* Cambridge: Cambridge University Press.

Leech, George

1981 *Semantics: The Study of Meaning.* Harmondsworth: Penguin.

Leitch, T. M.

1986 *What Stories Are: Narrative Theory and Interpretation.* University Park: Pennsylvania State University Press.

Lekomcev, J. K.

1977 Foundations of general semiotics. In: D. P. Lucid (ed.), *Soviet Semiotics*, 39-44. Baltimore: The Johns Hopkins University Press.

Lenneberg, Eric

1967 *The Biological Foundations of Language.* New York: John Wiley.

Levelt, W. J. M.

1989 *Speaking: From Intention to Articulation.* Cambridge: MIT Press.

Lévi-Strauss, Claude

1958 *Anthropologie structurale.* Paris: Librairie Plon.

1962 *Le totémisme aujourd'hui.* Paris: Presses Universitaires de France.

1964 *The Raw and the Cooked.* London: Cape.

Levin, Samuel

1977 *The Semantics of Metaphor.* Baltimore: Johns Hopkins Press.

1988 *Metaphoric Worlds.* New Haven: Yale University Press.

Lewis, C. I.

1946 *An Analysis of Knowledge and Valuation.* LaSalle: Open Court.

Leyhausen, P.

1967 Biologie von Ausdruck und Eindruck. *Psychologische Forschung* 31: 177-227.

Lieberman, Phillip

1972 *The Speech of Primates.* The Hague: Mouton.

1975 *On the Origins of Language.* New York: MacMillan.

1984 *The Biology and Evolution of Language.* Cambridge, Mass.: Harvard University Press.

1991 *Uniquely Human: The Evolution of Speech, Thought, and Selfless Behavior.* Cambridge, Mass.: Harvard University Press.

Liebman, R., S. Minuchin, and L. Baker

1974a An integrated program for anorexia nervosa. *American Journal of Psychiatry* 131: 432-435.

1974b The role of family in the treatment of anorexia nervosa. *Journal of the Academy of Child Psychology* 3: 264-274.

Linden, E.

1986 *Silent Partners: The Legacy of the Ape Language Experiments.* New York: Signet.

Liszka, James J.

1989 *The Semiotic Study of Myth: A Critical Study of the Symbol.* Bloomington: Indiana University Press.

Lloyd, J. E.

1966 *Studies on the Flash Communication System in Photinus Fireflies.* Ann Arbor: Museum of Zoology, University of Michigan.

Locke, John

1690 [1975] *An Essay Concerning Human Understanding*, P. H. Nidditch (ed.). Oxford: Clarendon Press.

Lorenz, Konrad

1952 *King Solomon's Ring.* New York: Crowell.

1971 *Studies in Animal and Human Behaviour.* Cambridge, Mass.: Harvard University Press.

Lotman, Juri

1977 Primary and secondary communication modeling systems. In: D. P. Lucid (ed.), *Soviet Semiotics*, 95-98. Baltimore: The Johns Hopkins University Press. 95-98.

1984 O Semiosfere. *Trudy po znakovym sistemam* 17: 5-623.

1994 Thesis towards a semiotics of Russian culture. *Elementa* 1(3): 219-227.

Lotman, Juri and B. A. Uspenskij (eds.)

1973 *Ricerche semiotiche.* Torino: Einaudi.

Lotman, Juri and B. A. Uspenskij

1978 On the semiotic mechanism of culture. *New Literary History* 9: 211-232.

Lott, D. F. and R. Sommer

1967 Seating arrangements and status. *Journal of Personality and Social Psychology* 7: 90-94.

Lucid, D. P. (ed.)

1977 *Soviet Semiotics: An Anthology*. Baltimore: The Johns Hopkins Press.

Lucy, John A.

1992 *Language Diversity and Thought: A Reformulation of the Linguistic Relativity Hypothesis*. Cambridge: Cambridge University Press.

Lumsden, C. J. and E. O. Wilson

1983 *Promethean Fire: Reflections on the Origin of Mind*. Cambridge, Mass.: Harvard University Press.

Luria, A. R.

1970 *Traumatic Aphasia*. New York: Humanities Press.

Lyons, John

1977 *Semantics*. Cambridge: Cambridge University Press.

MacCormac, Earl R.

1976 *Metaphor and Myth in Science and Religion*. Durham, N. C.: Duke University Press.

1985 *A cognitive theory of metaphor*. Cambridge, Mass.: MIT Press.

Mackay, A. L.

1984 The code breakers. *The Sciences* 24: 13-14.

Majno, G.

1975 *The Healing Hand*. Cambridge, Mass.: Harvard University Press.

Mallory, J. P.

1989 *In Search of the Indo-Europeans: Language, Archaeology and Myth*. London: Thames and Hudson.

Malson, L.

1973 Un entretien avec Claude Lévi-Strauss. *Le Monde* 20: 3-5.

Maritain, Jacques

1943 *Sign and Symbol: Redeeming the Time*. London: Geoffrey Bles.

1957 Language and the theory of sign. In: R. Nanda Anshen (ed.), *Language: An Enquiry into Its Meaning and Function*, 85-101. New York: Harper & Brothers.

Markus, R. A.

1957 St. Augustine on signs. *Phronesis* 2: 60-83.

May, Rollin

1991 *The Cry for Myth.* New York: Norton.

McBryde, C. M. and R. S. Backlow

1970 *Signs and Symptoms: Applied Pathologic Physiology and Clinical Interpretation*. Philadelphia: Lippincott.

McKean, K.

1982 Diagnosis by computer. *Discovery* 3: 62-65.

McLennan, J. F.

1869 The worship of animals and plants. *Fortnightly Review* 12: 407-427, 562-582.

McLuhan, Marshall

1962 *The Gutenberg Galaxy*. Toronto: University of Toronto Press.

1964 *Understanding Media.* London: Routledge & Kegan Paul.

McNeill, David

1987 *Psycholinguistics: A New Approach*. New York: Harper & Row.

1992 *Hand and Mind: What Gestures Reveal about Thought.* Chicago: University of Chicago Press.

Meiland, J. W.

1970 *The Nature of Intention*. London: Methuen.

Melzack, Robert

1972 The perception of pain. In: R. F. Thompson (ed.), *Physiological Psychology*, 223-231. San Francisco: Freeman.

1988 Pain. In: J. Kuper (ed.), *A Lexicon of Psychology, Psychiatry and Psychoanalysis*, 288-291. London: Routledge.

Metz, Christian

1974 *Film Language: A Semiotics of the Cinema.* Oxford: Oxford University Press.

1985 Photography and fetish. *October* 34: 81-90.

Miller, George A. and P. M. Gildea

1991 How children learn words. In: W. S-Y Wang (ed.), *The Emergence of Language: Development and Evolution*, 150-158. New York: W. H. Freeman.

Miller, J.

1978 *The Body in Question*. New York: Random House.

Mitchell, M.

1993 *Analogy-Making as Perception: A Computer Model.* Cambridge, Mass.: MIT Press.

Moenssens, A. A.

1971 *Fingerprint Techniques.* Philadelphia: Chilton.

Money, John

1986 *Lovemaps: Clinical Concepts of Sexual/Erotic Health and Pathology, Paraphilia, and Gender Identity from Conception to Maturity.* Baltimore: Johns Hopkins.

Morgan, C. L.

1895 *Introduction to Comparative Psychology.* London: Scott.

Morris, Charles W.

1946 *Signs, Language and Behavior.* New York: Prentice-Hall.

1938 *Foundations of the Theory of Signs.* Chicago: University of Chicago Press.

1946 *Signs, Language and Behavior.* Englewood Cliffs, N. J.: Prentice-Hall.

1971 *Writings on the General Theory of Signs.* The Hague: Mouton.

Morris, Desmond

1969 *The Human Zoo.* New York: McGraw-Hill.

1994 *The Human Animal.* London: BBC Books.

Morris, Desmond et al.

1979 *Gestures: Their Origins and Distributions.* London: Cape.

Mortenson, J.

1987 *Whale Song and Wasp Maps: The Mystery of Animal Thinking.* New York: Dutton.

Mounin, Georges

1981 Sémiologie médicale et sémiologie linguistique. *Confrontations Psychiatriques* 19: 43-58.

1970 *Introduction à la semiologie.* Paris: Les Editions de Minuit.

Müller, F. M.

1861 *Lectures on the Science of Language.* London: Longmans.

Munn, N. D.

1973 *Walbiri Iconography: Graphic Representation and Cultural Symbolism in a Central Australian Society.* Ithaca, NY: Cornell University Press.

Nelson, Katharine

1996 *Language in Cognitive Development: The Emergence of the Mediated Mind.* Cambridge: Cambridge University Press.

Nespoulous, J. L., P. Perron, and A. R. Lecours (eds.)

1986 *The Biological Foundations of Gestures: Motor and Semiotic Aspects.* Hillsdale, N. J.: Lawrence Erlbaum.

Neuburger, M.

1906 *Geschichte der Medizin.* Stuttgart: Enke.

Noiré, L.

1917 *The Origin and Philosophy of Language.* Chicago: Open Court.

Nöth, Winfred

1990 *Handbook of Semiotics.* Bloomington: Indiana University Press.

Ogden, C. K. and Richards, I. A.

1923 *The Meaning of Meaning.* New York: Harcourt, Brace.

Ong, Walter J.

1977 *Interfaces of the Word: Studies in the Evolution of Consciousness and Culture.* Ithaca: Cornell University Press.

Opie, I. and P. Opie

1959 *The Lore and Language of Schoolchildren.* Frogmore: Paladin.

Ortony, Andrew (ed.)

1979 *Metaphor and Thought.* Cambridge: Cambridge University Press.

Osgood, Charles E. and Thomas A. Sebeok (eds.)

1954 *Psycholinguistics: A Survey of Theory and Research Problems.* Bloomington: Indiana University Press.

Osgood, Charles E. and G. J. Suci

1953 Factor analysis of meaning. *Journal of Experimental Psychology* 49: 325-328.

Osgood, Charles E., G. J. Suci, and P. H. Tannenbaum

1957 *The Measurement of Meaning.* Urbana: University of Illinois Press.

Osolsobě, I.

1979 On ostensive communication. *Studia Semiotyczne* 9: 63-75.

Ostwald, P. F.

1968 Symptoms, diagnosis and concepts of disease: Some comments on the semiotics of patient-physician communication. *SocSeil* 7: 95-106.

Paget, Richard

1930 *Human Speech.* London: Kegan Paul.

Paine, R. and W. Sherman

1970 Arterial Hypertension. In: C. M. Macbryde and R. S. Backlow (eds.), *Signs and Symptoms*, 45-56. Philadelphia: Lippincott.

Park, David

1997 *The Fire within the Eye: A Historical Essay on the Nature and Meaning of Light.* Princeton: Princeton University Press.

Patterson, F. G.

1978 The gestures of a gorilla: Language acquisition in another pongid. *Brain and Language* 5: 72-97.

Patterson, F. G. and E. Linden

1981 *The Education of Koko.* New York: Holt, Rinehart and Winston.

Pavlov, Ivan

1902 *The Work of Digestive Glands.* London: Griffin.

Pazukhin, R.

1972 The concept of signal. *Lingua Posn* 16: 25-43.

Pedersen, H.

1931 *The Discovery of Language*. Bloomington: Indiana University Press.

Peirce, Charles S.

1868 Some consequences of four incapabilities. *Journal of Speculative Philosophy* 2: 140-151.

1931-1958 *Collected Papers of Charles Sanders Peirce*, Vols. I-VIII, C. Hartshorne and P. Weiss (eds.). Cambridge, Mass.: Harvard University Press.

Phillips, E. D.

1973 *Greek Medicine*. London: Thames and Hudson.

Piaget Jean

1969 *The Child's Conception of the World.* Totowa, N. J.: Littlefield, Adams & Co.

Piaget, Jean and J. Inhelder

1969 *The Psychology of the Child.* New York: Basic Books.

Pike, Kenneth

1967 *Language in Relation to a Unified Theory of the Structure of Human Behavior.* The Hague: Mouton.

Pitts, W. and W. S. McCulloch

1947 How we know universals: The perception of auditory and visual forms. *Bulletin of Mathematical Biophysics* 9: 127-149.

Pohl, J.

1968 *Symboles et langages*. Paris: Sodi.

Pollio, H., J. Barlow, H. Fine, and M. Pollio

1977 *The Poetics of Growth: Figurative Language in Psychology, Psychotherapy, and Education*. Hillsdale, N. J.: Lawrence Erlbaum Associates.

Polunin, I.

1977 The body as an indicator of health and disease. In: J. Blacking (ed.), *The Anthropology of the Body*, 34-56. London: Academic.

Popper, Karl

1972 *Objective Knowledge: An Evolutionary Approach.* Oxford: Clarendon.

1976 *The Unending Quest.* Glasgow: Harper Collins.

Popper, Karl and John Eccles

1977 *The Self and the Brain.* Berlin: Springer.

Premack, A.

1976 *Why Chimps Can Read.* New York: Harper and Row.

Premack, D. and A. J. Premack

1983 *The Mind of an Ape.* New York: Norton.

Preziosi, Donald

1979 *The Semiotics of the Built Environment: An Introduction to Architectonic Analysis.* Bloomington: Indiana University Press.

1989 *Rethinking Art History: Meditations on a Coy Science.* New Haven: Yale University Press.

Prieto, L. J.

1975 *Etudes de linguistique et de semiologie générales*. Geneva: Librairie Droz.

Prince, G.

1982 *Narratology: The Form and Functioning of Narrative*. Berlin: Mouton.

Prodi, G.

1981 Sintomo/diagnosi. *Ricerca-Socializzazione* 12: 972-992.

Propp, Vladimir J.

1928 *Morphology of the Folktale.* Austin: University of Texas Press.

Putnam, H.

1973 Meaning and Reference. *The Journal of Philosophy* 70: 699-711.

Raffler-Engel, Walburga von, Jan Wind, and A. Jonker (eds.)

1989 *Studies in Language Origins*. Amsterdam: John Benjamins.

Ransdell, J.

1986 Index. In: Thomas A. Sebeok (ed.), *Encyclopedic Dictionary of Semiotics* 1: 340-341.

Rector, Monica and Trinta, A. R.

1985 *Comunicação não verbal: A gestualidade Brazileira.* Petropolis: Editor Vozes.

Reichenbach, Hans

1948 *Elements of Symbolic Logic*. New York: Macmillan.

Révész, G.

1956 *The Origins and Prehistory of Language.* New York: Philosophical Library.

Revzina, O. G.

1972 The Fourth Summer School on Secondary Modeling Systems. *Semiotica* 6: 222-243.

Richards, I. A.

1936 *The Philosophy of Rhetoric.* Oxford: Oxford University Press.

1969 Tipi e campioni. *Strumenti Critici* 3: 187-193.

Roberts, D.

1973 *The Existential Graphs of Charles S. Peirce*. The Hague: Mouton.

Rohter, L.

1932 Macabre relic is laid to rest by Mexicans. *New York Times*, December 10, p. 9.

Roiphe, H.

1973 The infantile fetish. *Psychoanalytic Study of the Child* 28: 147-166.

Rosch, Elinor

1973a On the internal structure of perceptual and semantic categories. In: T. E. Moore (ed.), *Cognitive Development and Acquisition of Language*, 111-144. New York: Academic.

1973b Natural categories. *Cognitive Psychology* 4: 328-350.

Rousseau, Jean Jacques

1966 *Essay on the Origin of Language*, J. H. Moran and A. Gode (trans.). Chicago: University of Chicago Press.

Rowell, T.

1972 *The Social Behaviour of Monkeys*. Harmondsworth: Penguin.

Royce, Ann P.

1977 *The Anthropology of Dance*. Bloomington: Indiana University Press.

Rudy, S.

1986 Semiotics in the USSR. In: Thomas A. Sebeok and Jean Umiker-Sebeok (eds.), *The Semiotic Sphere*, 34-67. New York: Plenum.

Ruesch, J.

1972 *Semiotic Approaches to Human Relations*. The Hague: Mouton.

Rumbaugh, Duane M.

1977 *Language Learning by Chimpanzee: The Lana Project*. New York: Academic.

Russell, Bertrand

1940 *An Inquiry into Meaning and Truth*, London: Allen and Unwin.

Ruwet, N.

1991 *Syntax and Human Experience*. Chicago: University of Chicago Press.

Saint-Martin, Fernande

1990 *Semiotics of Visual Language*. Bloomington: Indiana University Press.

Sakitt, B.

1975 Locus of short-term visual storage. *Science* 190: 1318-1319.

Sanders, G.

1970 Peirce's sixty-six signs? *Transactions of the Charles S. Peirce Society* 6: 3-16.

Sapir, Edward

1921 *Language*. New York: Harcourt, Brace, and World.

1929 The status of linguistics as a science. *Language* 5: 207-214.

1931 Communication. *Encyclopedia of the Social Sciences* 4: 78-481.

Sarton, G.

1954 *Galen of Pergamon*. Lawrence: University of Kansas Press.

Saussure, Ferdinand de

1916 *Cours de linguistique générale*. Paris: Payot.

Savage-Rumbaugh, E. S., D. M. Rumbaugh, and S. L. Boysen

1978 Symbolic communication between two chimpanzees. *Science* 201: 641-644.

Savan, David

1983 Toward a refutation of semiotic idealism. *Semiotic Inquiry* 3: 1-8.

Sayers, D. L.

1932 *Have His Carcase*. New York: Harcourt, Brace and Company.

Schindler, W.

1953 A case of crutch fetishism as the result of a literal Oedipus complex. *International Journal of Sexology* 6: 131-135.

Schleidt, M.

1980 Personal odor and nonverbal communication. *Ethology and Sociobiology* 1: 225-231.

Schmandt-Besserat, D.

1978 The earliest precursor of writing. *Scientific American* 238: 50-59.

1989 Two precursors of writing: Plain and complex tokens. In: W. M. Senner (ed.), *The Origins of Writing*, 27-40. Lincoln: University of Nebraska Press.

1992 *Before Writing*, 2 vols. Austin: University of Texas Press.

Schneirla, T. C.

1965 Aspects of stimulation and organization in approach/withdrawal processes underlying vertebrate behavioral development. *Advances in the Study of Behavior* 1: 1-74.

Schogt, Henry

1988 *Linguistics, Literary Analysis, and Literary Translation*. Toronto: University of Toronto Press.

Scholes, Robert

1982 *Semiotics and Interpretation.* New Haven: Yale University Press.

Schor, N.

1985 Female fetishism: The case of George Sand. *Poetics Today* 6: 301-310.

Searle, John R.

1984 *Minds, Brain, and Science*. Cambridge, Mass.: Harvard University Press.

Sebeok, Thomas A.

1963 Communication among social bees; porpoises and sonar, man and dolphin. *Language* 39: 448-466.

1968 *Animal Communication: Techniques of Study and Results of Research*. Bloomington: Indiana University Press.

1972a *Perspectives in Zoosemiotics*. The Hague: Mouton.

1972b Problems in the classification of signs. In: *Studies for Einar Haugen*, 511-521. The Hague: Mouton.

1973a Semiotics: A survey of the state of the art. In: Thomas A. Sebeok (ed.), *Current Trends in Linguistics 12*, 161-213. The Hague: Mouton.

1973b Semiotica e affini. *VS: Quaderni di Studi Semiotici* 3: 1-11.

1976 *Contributions to the Doctrine of Signs*. Lisse: The Netherlands.

1979 *The Sign and Its Masters*. Austin: University of Texas Press.

1981a *The Play of Musement*. Bloomington: Indiana University Press.

1981b Karl Bühler. In: Martin Krampen et al. (eds.), *Die Welt als Zeichen: Klassiker der modernen Semiotik*, 34-46. Berlin: Severin und Siedler.

1985 *Contributions to the Doctrine of Signs*. Lanham: University Press of America.

1985 Pandora's box: How and why to communicate 10,000 years into the future. In: M. Blonsky (ed.), *On Signs*, 448-466. Baltimore: Johns Hopkins University Press.

1989 Fetish. *American Journal of Semiotics* 6: 51-65.

1990 *Essays in Zoosemiotics*. Toronto: Toronto Semiotic Circle.

1991a *A Sign is Just a Sign*. Bloomington: Indiana University Press.

1991b *Semiotics in the United States*. Bloomington: Indiana University Press.

1994 *Signs: An Introduction to Semiotics*. Toronto: University of Toronto Press.

Sebeok, Thomas A. and Jean Umiker-Sebeok (eds.)

1992 *Biosemiotics*. Berlin: Mouton de Gruyter.

Shands, H. C.

1970 *Semiotic Approaches to Psychiatry*. The Hague: Mouton.

Shannon, Claude E.

1948 A mathematical theory of communication. *Bell Systems Technical Journal* 27: 379-423.

Shannon, Claude E. and Warren Weaver

1949 *The Mathematical Theory of Communication*. Urbana: University of Illinois Press.

Shapiro, Michael

1983 *The Sense of Grammar: Language as Semeiotic*. Bloomington: Indiana University Press.

Sherzer, J.

1973 Verbal and nonverbal deixis: The pointed lip gesture among the San Blas Cuna. *Language in Society* 2: 117-131.

Shevoroshkin, V. (ed.)

1989 *Reconstructing Languages and Cultures.* Bochum: Brockmeyer.

Short, T. L.

1982 Life among the legisigns. *Transactions of the Charles S. Peirce Society* 18: 285-310.

Siegel, R. E.

1973 *Galen on Psychology, Psychopathology, and Function and Diseases of the Nervous System.* Basel: Karger.

Silverman, Kaja

1983 *The Subject of Semiotics.* Oxford: Oxford University Press.

Skinner, B. F.

1938 *The Behavior of Organisms.* New York: Appleton-Century-Crofts.

Skousen, R.

1989 *Analogical Modeling of Language.* Dordrecht: Kluwer.

Smith, C. G.

1985 *Ancestral Voices: Language and the Evolution of Human Consciousness.* Englewood Cliffs, N. J.: Prentice-Hall.

Smith, W. John

1965 Message, meaning, and context in ethology. *The American Naturalist* 99: 405-409.

1969a Displays and messages in intraspecific communication. *Semiotica* 1: 357-369.

1969b Messages of vertebrate communication. *Science* 165: 145-150.

1977 *The Behavior of Communicating: An Ethological Approach.* Cambridge, Mass.: Harvard University Press.

Sonnesson, G.

1989 *Pictorial Concepts: Inquiries into the Semiotic Heritage and Its Relevance for the Analysis of the Visual World.* Lund: Lund University Press.

Sontag, Susan

1978 *Illness as Metaphor.* New York: Farrar, Straus & Giroux.

1989 *AIDS and Its Metaphors.* New York: Farrar, Straus & Giroux.

Sørensen, H. S.

1963 *The Meaning of Proper Names.* Copenhagen: Gad.

Sperling, M.

1963 Fetishism in children. *Psychoanalytic Quarterly* 32: 374-392.

Spang-Hanssen, Henning

1954 Recent theories on the nature of the language sign. *TCLC* , whole volume 9.

Staal, J. F.

1971 What was left of pragmatism in Jerusalem. *Language Sciences* 14: 29-32.

Staehlin, W.

1914 Zür die Psychologie und Statistik der Metapher. *Archiv für Gesamte Psychologie* 31, 299-425.

Stahl, S.

1989 *Literary Folkloristics and the Personal Narrative*. Bloomington: Indiana University Press.

Staiano, K. V.

1979 A semiotic definition of illness. *Semiotica* 28: 107-125.

1982 Medical semiotics: Redefining an ancient craft. *Semiotica* 38: 319-346.

Stamp Dawkins, M.

1993 *The Search for Animal Consciousness*. Oxford: Freeman.

Stanosz, B.

1970 Formal theories of extension and intension of expressions. *Semiotica* 2: 102-114.

Stewart, I.

1975 The seven elementary catastrophes. *New Scientist* 68: 447-454.

Stratton, J.

1987 *The Virgin Text: Fiction, Sexuality, and Ideology*. Norman: University of Oklahoma Press.

Stross, Brian

1976 *The Origin and Evolution of Language*. Dubuque, Iowa: W. C. Brown.

Sturtevant, E. H.

1947 *An Introduction to Linguistic Science*. New Haven: Yale University Press.

Swadesh, Morris

1951 Diffusional cumulation and archaic residue as historical explanations. *Southwestern Journal of Anthropology* 7: 1-21.

1959 Linguistics as an instrument of prehistory. *Southwestern Journal of Anthropology* 15: 20-35.

1971 *The Origins and Diversification of Language.* Chicago: Aldine-Atherton.

Sweetser, Eve

1990 *From Etymology to Pragmatics: The Mind-as-Body Metaphor in Semantic Structure and Semantic Change.* Cambridge: Cambridge University Press.

Taylor, John R.

1995 *Linguistic Categorization: Prototypes in Linguistic Theory.* Oxford: Oxford University Press.

Telegdi, Z.

1976 Zur Herausbildung des Begriffs "sprachliches Zeichen" und zur stoischen Sprachlehre. *Acta Linguistica Scientiarum Hungaricae* 26: 267-305.

Tembrock, G.

1971 *Biokommunikation: Informationsbetragung im biologischen Bereich.* Berlin: Akademie-Verlag.

Temkin, O.

1973 *Galenism.* Ithaca: Cornell University Press.

Terrace, H. S.

1979 *Nim.* New York: Knopf.

Thibaud, Paul

1975 *La logique de Charles Sanders Peirce: De l'algèbre aux graphes.* Aix-en-Provence: Université de Provence.

Thom, René

1973 De l'icône au symbole: Esquisse d'une théorie du symbolisme. *Cahiers Internationaux de Symbolisme* 22-23: 85-106.

1974 La linguistique, discipline morphologique exemplaire. *Critique* 30: 235-245.

1975 *Structural Stability and Morphogenesis: An Outline of a General Theory of Models.* Reading: W. A. Benjamin.

1980 L'espace et les signes. *Semiotica* 38: 205-215.

Thorndyke, E. L.

1898 *Animal Intelligence.* New York: Macmillan.

Thorpe, W. H.

1961 *Bird-song.* Cambridge: Cambridge University Press.

1967 Vocal imitation and antiphonal song and its implications. In: D. W. Snow (ed.), *Proceedings of the XIV International Ornithological Congress*, 245-263. Oxford: Blackwell.

Tinbergen, Nikolaas

1963 On aims and methods of ethology. *Zeitschrift für Tierpsychologie* 20: 410-433.

Tinbergen, Nikolaas and A. C. Perdeck

1950 On the stimulus situation releasing the begging response in the newly hatched herring gull chick. *Behaviour* 3: 1-39.

Todorov, Tvetzan

1973 Semiotics. *Screen* 14: 15-23.

Toller, S. van and G. H. Donn (eds.)

1989 *Perfumery: The Psychology and Biology of Fragrance*. New York: Routledge, Chapman and Hall.

Toolan, M. J

1988 *Narrative: A Critical Linguistic Introduction.* London: Routledge.

Trabant, Jürgen

1981 *Die Welt als Zeichen. Klassiker der modernen Semiotik.* Berlin: Severin and Siedler.

Trevarthen, Colin

1989 Signs before speech. In: Thomas A. Sebeok and Jean Umiker-Sebeok (eds.), *The Semiotic Web 1988*, 689-755. Berlin: Mouton de Gruyter.

Tsuda, A.

1984 *Sales Talk in Japan and the United States. An Ethnographic Analysis of Contrastive Speech Event* Washington D. C.: Georgetown University Press.

Tubbs, Stewart L. and Silvia Moss

1991 *Human Communication*. New Yorl: McGraw-Hill.

Tufte, E. R.

1997 *Visual Explanations: Images and Quantities, Evidence and Narrative.* Cheshire: Graphics Press.

Tulving, E.

1972 Episodic and semantic memory. In: E. Tulving and W. Donaldson (eds.), *Organization of Memory*, 23-46. New York: Academic.

Uexküll, Jakob von

1909 *Umwelt und Innenwelt der Tiere*. Berlin: Springer.

1928 *Theoretische Biologie.* Frankfurt: Suhrkamp.

1982a The theory of meaning. *Semiotica* 42: 1-87.

Uexküll, Thure von

1982b Semiotics and medicine. *Semiotica* 38: 205-215.

1989 Jakob von Uexküll's Umwelt-theory. In: Thomas A. Sebeok and Jean Umiker-Sebeok (eds.), *The Semiotic Web 1988*, 129-158. Berlin: Mouton de Gruyter.

Uexküll, Thure von et al.

1979 *Lehrbuch der Psychosomatischen Medizin.* München: Urban & Schwarzenberg.

Uexküll, Theure von (ed.)

1981 *Kompositionslehre der Natur: Biologie als undogmatische Naturwissenschaft by Jakob von Uexküll.* Frankfurt am Main: Verlag Ullstein Propyläen.

Ullmann, S.

1951 *Principles of Semantics*. Glasgow: Jackson, Son & Co.

Uspenskij, Boris A.

1991 *Semiotik der Geschichte*. Wein: Österreichische Akademie der Wissenschaften.

Valesio, Paolo

1969 Icons and patterns in the structure of language. *ACII* 10: 383-387.

Van Wing, R. P. J.

1938 *Etudes bakongo II, Religion et magie*. Brussels: G. van Campenhout.

Varela, F. J., E. Thompson, E., and E. Rosch

1991 *The Embodied Mind: Cognitive Science and Human Experience*. Cambridge, Mass.: MIT Press.

Viberg, Ann

1983 The verbs of perception: A typological study. *Linguistics* 21: 123-162.

Vigener, G.

1989 Dieser Schuch ist kein Schuch—Zur Semiotik des Fetischs. In: J. Bernard, T. Klugsberger, and G. Wiltham (eds.), *Semiotik der Geschlechter. Akten des 6. Symposiums der österreichischen Gesellschaft für Semiotik, Salzburg 1987*, 341-352. Stuttgart: Heinz.

Vygotsky, Lev S.

1931 *Storia dello sviluppo delle funzioni psichiche superiori.* Firenze: Giunti-Barbèra.

1962 *Thought and Language.* Cambridge, Mass.: MIT Press.

1978 *Mind in Society.* Cambridge, Mass.: Harvard University Press.

Wallis, M.

1973 On iconic signs. In: J. Rey-Debove (ed.), *Recherches sur les systèmes signifiants*, 481-498. The Hague: Mouton.

Walther, E.

1984 Die Beziehung zwischen Semiotik und Linguistik. *Semiotica* 52: 111-117.

Ward, L. and Walburga von Raffler-Engel

1980 The impact of nonverbal behavior on foreign language teaching. In: Walburga von Raffler-Engel (ed.), *Aspects of Nonverbal Communication*, 287-304. Lisse: Swets and Zeitlinger.

Watson, James and Anne Hill

1993 *A Dictionary of Communication and Media Studies*. London: Edward Arnold.

Watson, John B.

1929 *Psychology from the Standpoint of a Behaviorist.* Philadelphia: Lippincott.

Watzlawick, Paul

1978 *The language of change*. New York: W. W. Norton.

Waugh, Linda R.

1982 Marked and unmarked: A choice between unequals in semiotic structure. *Semiotica* 38: 299-318.

Way, E. C.

1991 *Knowledge Representation and Metaphor.* Dordrecht: Kluwer.

Weckler, W.

1973 *The Sexual Code*. New York: Anchor.

Weimann, W.

1962 Über Tätowierungsfetischismus. *Archiv für Kriminologie* 130: 106-109.

Weinreich, Ulrich

1968 Semantics and semiotics. *International Encyclopedia of the Social Sciences* 14: 164-69.

Weiss, P. and A. Burks

1945 Peirce's sixty-six signs. *Journal of Philosophy* 42: 383-388.

Wells, Gordon

1986 *The Meaning Makers: Children Learning Language and Using Language to Learn.* Portsmouth: Heinemann.

Wells, Rulon

1954 Meaning and use. *Word* 10: 235-250.

1967 Distinctively human semiotic. *Social Science Information* 6: 103-124.

Welte, W.

1974 *Moderne Linguistik*. Munich: Max Hueber.

Werner, H. and B. Kaplan

1963 *Symbol Formation: An Organismic-Developmental Approach to the Psychology of Language and the Expression of Thought*. New York: John Wiley.

Wertheimer, M.

1923 Untersuchungen zur Lehre von der Gestalt, II. *Psychologische Forschungen* 4: 301-350.

Wescott, Roger W.

1971 Linguistic iconism. *Language* 47: 416-428.

1978 Visualizing vision. In: B. Randhawa and W. Coffman (eds.), *Visual Learning, Thinking, and Communication*, 21-37. New York: Academic Press.

Wescott, Roger W. (ed.)

1974 *Language Origins*. Silver Springs, Maryland: Linstok Press.

Wheeler, J. A.

1988 World as system self-synthesized by quantum networking. *IBM Journal of Research and Development* 32: 1-15.

Wheelwright, P.

1954 *The Burning Fountain: A Study in the Language of Symbolism*. Bloomington: Indiana University Press.

White, L. A.

1940 The symbol: The origin and basis of human behavior. *Philosophy of Science* 7: 451-463.

Whorf, Benjamin Lee

1956 *Language, Thought, and Reality*, J. B. Carroll (ed.). Cambridge, Mass.: MIT Press.

Wickler, W.

1968 *Mimicry in Plants and Animals*. New York: McGraw Hill.

Wiener, Norbert

1949 *Cybernetics, or Control and Communication in the Animal and the Machine*. Cambridge, Mass.: MIT Press.

Wierzbicka, Ann

1996 *Semantics: Primes and Universals*. Oxford: Oxford University Press.

Wilder, H. H. and B. Wentworth

1918 *Personal Identification*. Boston: Badger.

Wills, D. D.

1990 Indexifiers in Wolof. *Semiotica* 78: 193-218.

Wilson, C.

1989 *The Misfits: A Study of Sexual Outsiders*. London: Carroll and Graf.

Wilson, Edward Osborne

1971 The prospects for a unified sociobiology. *American Scientist* 59: 400-403.

1975 *Sociobiology: The New Synthesis.* Cambridge, Mass.: Harvard University Press.

1979 *On Human Nature.* New York: Bantam.

1984 *Biophilia.* Cambridge, Mass.: Harvard University Press.

Wilson, Edward Osborne and M. Harris

1981 Heredity versus culture: A debate. In: J. Guillemin (ed.), *Anthropological Realities: Reading in the Science of Culture*, 459-467. New Brunswick, N. J.: Transaction Books.

Wimsatt, W. R.

1954 *The Verbal Icon: Studies in the Meaning of Poetry.* New York: University of Kentucky Press.

Wintsch, S.

1979 The vocabulary of gestures: Nonverbal communication in foreign languages. *Research & Creative Activity* 3: 6-11.

Yerkes, Robert

1916 *The Mental Life of Monkeys and Apes*. New Haven: Yale University Press.

Zavitzianos, G.

1971 Fetishism and exhibitionism in the female and their relationship to psychopathy and kleptomania. *International Journal of Psycho-Analysis* 52: 297-305.

Zeman, J. J.

1964 *The Graphical Logic of C. S. Peirce*. Dissertation, University of Chicago.

Index